New Developments in Medical Research

Gut Bacteria

The Secret to Health

NEW DEVELOPMENTS IN MEDICAL RESEARCH

Additional books and e-books in this series can be found on Nova's website under the Series tab.

NEW DEVELOPMENTS IN MEDICAL RESEARCH

GUT BACTERIA

THE SECRET TO HEALTH

UJJWAL SONIKA, MD, DM

AND

MEDHA KAPOOR, PHD

Library of Congress Cataloging-in-Publication Data

ISBN: 978-1-53619-084-7

Published by Nova Science Publishers, Inc. † New York

To Oeshi for always filling my heart with joy

U.S.

To my daughter Aadya who lights up my world with her joyful laughter and unconditional love.

M.K.

CONTENTS

PREFACE

We live in challenging times, each of us braving the impact of the coronavirus pandemic on our lives. The illness caused by SARS CoV-2, COVID-19 is usually mild but can be severe and life-threatening in few, particularly those with comorbidities. Most of these comorbidities are lifestyle diseases like diabetes, hypertension, and obesity. The idea to write this book however, was conceived in the pre-covid era. Our first encounters with the gut bacteria occurred while reading various research articles and scientific studies in our respective specialties. That was when we realized the importance of these tiny creatures and our ignorance about them. What followed was a reading frenzy about the gut microbiome and its role in our health and various diseases. The research related to the gut microbiome is progressing at a tremendous pace, and not many people, even in the fields of health and nutrition sciences are aware of the latest developments. Besides, the impact of gut bacteria on us is so powerful and widespread that it is essential to know about them for everyone who wishes to lead a healthy living. It is this thought which is the guiding light behind this book.

Our most crucial endeavor has been to provide the latest scientific information in an easy-to-understand language to the benefit of everyone who reads it, irrespective of their core background. We have gone through more than 800 journal articles and aimed to distill the knowledge in the form of this book which you now hold. The book is structured so that the reader gets to know about the gut-flora, its functions in health, and its role in various lifestyle diseases. And finally, you get to know the ways to achieve a healthy gut-flora to attain optimal health. The book uses figures and illustrations for easy understanding of relatively complex concepts. A few essential topics other than the gut microbiome have also been discussed that directly or indirectly relate to gut health. We have deliberately refrained from including the name of bacterial species in the book to keep the message as crisp as possible for each reader, irrespective of the academic background. On occasions, it has been absolutely necessary to name the bacterial species, but one should not worry about remembering them. After reading this book, we hope that you will look at bacteria in a new light as friends who confer many health benefits and use that knowledge to maximize your own health and well-being. If this happens, it will give us immense satisfaction.

Ujjwal Sonika
Medha Kapoor

INTRODUCTION

What do you think when you come across the word 'bacteria?' Do you think of infections, diseases, illness and antiseptics or friends, health, and probiotics? Most of us believe that bacteria are our enemies. They cause infections, diseases, illnesses, and can only harm us. The repeated conversations about hospital-acquired infections, and antimicrobial resistance makes us all the more fearful. And we seem to be engaged in a constant battle against these tiny creatures. But the truth is that bacteria are amongst our oldest friends. Not just are they everywhere in the environment constantly surrounding us, they are also present 'on' as well as 'inside' us. Out of all the bacteria that constitute our world, the ones residing in our intestines are the most special.

What are these bacteria?

How were they discovered?

What are the functions they perform in our body?

As you read through this book, you will get to know the answers to all these questions and appreciate the role of bacteria in regulating our health and well-being. What would be particularly intriguing is understanding the tremendous role these tiny creatures play in illnesses such as diabetes and obesity, also known as the lifestyle diseases. Not only this, the gut bacteria even impact our mood and behavior. They influence not only our physical health but also our mental and emotional status. The bacteria in our gut can communicate with our brain and influence its functioning.

The gut bacteria live in a state of delicate balance. A large number of factors determine the constitution of our gut-flora. A multitude of factors can incline the gut-flora composition towards an unhealthy state, which may result in diseases. On the bright side, this also means that by adopting healthy lifestyle practices, we can improve our gut bacteria to achieve good health. As the gut bacteria directly impact our well-being, a healthy gut can lead to a healthy body and a sound mind. Although we shall learn about how healthy gut-flora looks, this knowledge is of no benefit without action. One must be able to put the knowledge acquired into an actionable plan. The ways to achieve a healthy gut are simple but require life-long persistence as gut bacteria are notorious for getting back to their original state. You can achieve a healthy gut by adopting healthy lifestyle practices without any fancy and expensive tools. Let us explore the book to discover the road to a healthy gut-flora and, consequently, a healthy and happy life!

ACKNOWLEDGMENTS

My sincere thanks to Dr. Medha for this book could not have been written without her. She has been a continuous source of energy and inspiration while writing this book.

I want to thank my wife, Kanika, and my daughter Oeshi for their support and patience. It was the time supposed to be spent with them that was utilized for writing the book.

I am especially thankful to my parents, my father, Mr. Shyam, my mother, Mrs. Renu, and my sister Surbhi who have supported me through every phase of my life.

I extend my gratitude towards my colleagues Prof. Sharma, Prof. Sachdeva, Prof. Srivastava, Dr. Kumar, and Dr. Dalal, for their continued support and encouragement while writing this book.

I am grateful to Dr. Apala, an expert psychiatrist, for her invaluable insights into stress, behavior, and sleep.

I thank all my esteemed teachers. It is their blessings that have made this work possible.

I especially thank Nova publications for believing in us and providing us this opportunity to publish with them.

Finally I thank you, the reader, for trusting our book and putting your time and money into it.

Ujjwal Sonika

Dr. Ujjwal Sonika and I have admired each other's professional work for the past few years and have earlier collaborated on a few projects. One fine day, it occurred to us, that we should amalgamate our professional expertise in the form of a book, highlighting the importance of gut microbiome and its role in various lifestyle diseases. Today, after having spent months compiling this book, I am overjoyed that we had that conversation. I have learned a lot in the process, and I am grateful to Dr. Sonika for trusting me with the book.

I thank my doting husband Pranjal, for being a constant pillar of support, due to which I was able to juggle my professional engagements, this book, and the responsibilities of the early months of daunting yet the most beautiful project that I have ever undertaken — motherhood. Thank you for everything.

I thank my darling daughter, my heartbeat, my love, my happiness, my sunshine — Aadya, for being such a delightful infant who cooperated with mum-mum-mum-mamma-aa-mum-mum-mum throughout the journey of writing this book. I love you!

I am especially thankful to my grandmother Ms. Shail Kumari, my father Drs. Naresh Kapoor and my mother, Pratibha who have supported me unconditionally through every phase of my life. It is said that only after becoming a parent does one understand what ones' parents undergo to raise them. Mumma and Papa, I couldn't agree more!

I extend my gratitude towards my parents-in-law - Drs. Satish Chandra Joshi and Anita Joshi for being supportive of my professional and personal endeavors and showering me with love and blessings.

I thank my wonderful friends for keeping me sane during the hardships and making me laugh, sometimes, a bit too hard, when I find myself in a tough spot. I love you guys! You are A-awesome!

Every idea needs people who believe in it and nurture it; I thank Nova Publications for the wonderful opportunity to publish this book.

Medha Kapoor

SECTION 1

GUT-FLORA: THE ULTIMATE INSIDER

Am I Me
Or
Am I We

Chapter 1

GUT-FLORA: WHAT IS IT?

We are continuously bombarded with commercials promoting antibacterial products ranging from antiseptic solutions, germicidal soaps, and cleaners, to nowadays even antibacterial paints and light bulbs!! All these products claim to kill the bacteria present on our body or the surfaces surrounding us and provide safety to us, our children, and the elders. We assume that these products safeguard our health and believe that we need to constantly kill the bacteria present all around us to remain free from diseases. But is it really true? Do we need to keep ourselves, our houses, and work-places bacteria-free? Do all bacteria harm us?

Let us go back and look at the beginning of our journey with bacteria. In1676 AD, when Dutch scientist Anton van Leeuwenhoek observed tiny creatures in water through a microscope, bacteria were first discovered. He called them *animalcules*, which means small animals. In 1838, the German scientist Christian Gottfried Ehrenberg gave them their name bacteria, which means 'little stick' in Greek as the earliest observed bacteria were shaped like rods. However, bacteria can also be round as well as spiral in shape. For around the next 200 years, humans did not know that these small creatures could cause a disease. In the mid-nineteenth century, we begin to learn about bacterial infections. A British physician John Snow investigated a cholera epidemic in London in 1854 and found that the outbreak source was a public water pump at Broad Street. In the same year, Italian anatomist Filippo Pacini_discovered *Vibrio cholerae,* the bacterium which causes cholera. Pacini performed autopsies on cholera patients in Florence, Italy, immediately after their death and discovered the comma-shaped bacteria in their intestines. But it was 30 years later, when a more famous scientist Robert Koch made similar discoveries, it was accepted that cholera is caused by *Vibrio Cholerae.* The work of these pioneers established that bacteria could cause disease. At around the same time in mid-1880s, it was also discovered that the bacteria could reside in our intestines as a normal flora without any causing any harm. An Austrian pediatrician Theodor Escherich observed rod-shaped bacteria in the stools of children. These bacteria were later named Escherichia coli, after him. However, the knowledge that a vast number and different types of bacteria live in our intestines and perform a wide range of essential functions for our body has been gained only in the last 12-15 years.

So, do all bacteria harm us? Well, not really. In fact, the truth is that trillions of bacteria inhabit our bodies. A number so huge it far exceeds the number of human cells. The human cells present in our body are only one-tenth of the number of microbes. Take a pause; you are only 10 percent human and 90 percent microbes. These microbes live on our skin, inside our nose, oral cavity, stomach, and the female genitalia. Besides, a large number of them reside in our intestines and help us sustain basic body functions. Not only their numbers are huge, but their impact is also great. They play a key role in not just our physical but also mental and emotional health. It is true that bacteria can harm us, but the number of such harmful bacteria is minuscule when compared to our gut inhabitants. If you insist on a count, not even 100 types of bacteria cause human diseases, and more than a thousand types are living inside us, day and night, from the time we are born till we die. They are our own. They are us. Or maybe we are theirs.

WHAT IS GUT-FLORA?

The food we eat passes through our stomach and reaches the intestines, where the complex substances are broken down into simpler forms, and then absorbed into the blood-stream. These provide energy for all our activities as well as the nutrients which help build and repair the tissues. Our intestines are divided into two parts: small intestine and large intestine. These are like hollow tubes through which the food passes and gets digested and absorbed. The small intestine mainly absorbs the nutrients from the food into the blood-stream, and the large intestine is where the water is absorbed. As we just learned our intestines are inhabited by a large number of bacteria and other microbes (single-celled organisms), collectively known as the 'gut microbiota' or the 'gut-flora'. Their genes are collectively called 'gut microbiome'.

Although gut-flora includes organisms other than bacteria such as viruses and fungi as well, the bacteria are the ones that are most extensively studied. Their role in maintaining normal physiological processes and in various disease states is well-established. Therefore, for the purpose of this book, gut-flora is the sum total of all the types of bacteria residing in our intestines.

GUT BACTERIA: FACTS

Our gut-flora weighs around a massive 2 Kg in an adult!

A healthy adult harbors around 1000 different bacterial species at a time in the gut. These different varieties of bacteria present in our intestines at any given time perform a range of vital functions. Similar to the variety, the number of bacteria present in our gut is enormous too. It is estimated that around 10^{12} to 10^{14} i.e., a whopping 100 trillion bacteria reside in the intestines of each of us. The colon (large intestine) contains approximately 10^9–10^{12} bacteria/ml followed by 10^4–10^8bacteria/ml in the ileum and10^1–10^3 bacteria/ml in jejunum[1].

For ease of understanding, let us consider our intestine as a country and bacteria as the people; the nation of the human intestine has a population of around 10^{12} to 10^{14} people with 500 -1000 different ethnicities, variably distributed across different regions - some with a higher population density (large intestine) and others with lower population density. Compare to the real world; the world's human population is around 7.8 x 10^9 (billion) people. Well, do you realize? The number of bacteria residing in the intestines of a single person is at least 1000 times the number of humans living on the entire earth!! Just imagine!! In-fact, the human gut-flora is said to be the most densely populated and the most varied ecosystem on our planet. No, it is not the Amazon forests that are most densely populated. It is your own gut.

The combined gene pool of an organism is termed as the *genome*. The Human Genome Project, which identified and sequenced all the human genes, was completed in 2003. It revealed that the human genome consists of about 23 000 genes. Could you guess that what is the sum total of the genes present in our gut-flora? You would be flabbergasted to know that the gut-flora encodes over three million genes. Yes! The combined gene pool of the bacteria present inside our intestines is more than 100 times our own gene pool. These bacterial genes carry information necessary for making thousands of metabolites, which are not produced by us and perform many vital functions in our body. If we look at the genetic level, we are less than 1 percent human. Can you say that you are an individual when you really are a colony of trillions of microorganisms?

What are the types of bacteria present in our gut-flora? The bacteria present in our intestines are mostly anaerobes and belong to two predominant phyla (groups); Firmicutes and Bacteroidetes. The other important phyla include Proteobacteria and Actinobacteria. However, most of the changes in gut-flora, which are associated with lifestyle diseases,

---

[1] Jejunum and ileum are part of the small intestine. The proximal part near to the stomach is the jejunum, and the distal part is the ileum.

correspond with an alteration in the number and diversity of the Firmicutes and the Bacteroidetes. The gut-flora exists in a state of delicate balance, and its composition is governed by an interplay between a large number of factors that we will learn about pretty soon!

Did you know that your gut microbial signature is unique? No other human being has the same gut-flora as you do. Who knows, probably in the future, the governments across the world might decide to maintain a database of our unique gut microbiota just like the fingerprints and retina scans! If anyone commits a crime, besides finger-prints, he might have to provide his poop for identification! Tacky yet effective evidence indeed! If you pick up a deck of cards (52 cards) and shuffle it, there are about 80 unvigintillion (80 followed by 66 zeros!!) combinations that are possible. The chances are almost 100% that the combination you come up with will be unique; the only one ever achieved! In all probability, no one would ever come up with the same sequence upon random shuffling of cards. You produced something that was unique in the history of the universe, and chances are it would never be repeated ever. On a philosophical note, the next time you find yourself feeling worthless, pick up a deck of cards and shuffle it and yes, congratulate yourself for achieving what none ever has! The card's example may appear frivolous, but it makes a point. You ARE special! The universe arranged each human cell to make you UNIQUE. With your signature gut-flora, you become all the more unique! The one and only one of its kind!

Kim Jong-un, the North Korean commander-in-chief, visited Singapore in June 2018 for a bilateral summit with US president, Donald Trump. It was his first official visit to a state other than China and South Korea. Large hopes were pinned on the meeting of two of the world's most maverick leaders. Why are you reading about this visit? We are not talking about nuclear de-escalation. Well! We are interested in a tiny insignificant but creepy detail of this visit. Kim Jong-un carried his own toilet to the summit. Why? Lee Yun-keol, a former member of the North Korean Guard Command unit, told the Washington Post that it was because no one could have the access to his stools. His poops carry the information about his health status, information which is too secretive to be left behind anywhere. Now, you believe us. Your doubts about the magical powers of your stools are fading away. It is fascinating to see how easily political theories make us believe something that a scientific argument cannot. One can extract so much information about anyone from his/her stools. And we wondered that only Google knows more about us than we know ourselves!

You learned that a large number of bacteria reside on as well as inside us. Only a few of them are harmful. And there is a particularly special group which resides in our intestines. Why is that these bacteria present in the intestines have a special significance? Bacteria, too, reside on the skin, nose, mouth, stomach, and inside the vagina. But what is about gut bacteria that people write books about them and expect others to read? The reason is the place they reside, our intestines. We understand intestines as only an organ of our digestive system where nutrients and water are absorbed. But they are much more. Our intestines are the largest immune organ of our body. They have their own nervous system and can impact even our brain. Gut bacteria not only have an effect on metabolic function by being at the site of absorption of energy, but they are also on the driver's seat as far as our immune system is concerned. Besides, they can be easily obtained for the purpose of the study. All of us excrete them daily and provide enough samples to the researchers without the need for any sophisticated tools or complicated procedures.

What Are the Factors Which Affect Gut-Flora?

Our gut-flora is unique, and its constituent bacteria exist in a state of delicate balance (figure 1). What makes it unique are our experiences, practices, and certain random events which can influence the gut bacteria. It is easy to guess that our diet is amongst the most important governors of the gut-flora as these bacteria are directly exposed to what we eat. However, you might be surprised to know that a seemingly unrelated event, the mode of birth profoundly influences the gut-flora, particularly during the early childhood. Even having an elder sibling has an impact. These are not the figments of our imagination; science says so (Laursen et al. 2015). Besides, a range of our daily routines and practices influence these bacteria. To optimally harness their goodness, one must have knowledge about the factors influencing them. Some of these factors, such as our diet, can be modified while others cannot. Let us learn about them.

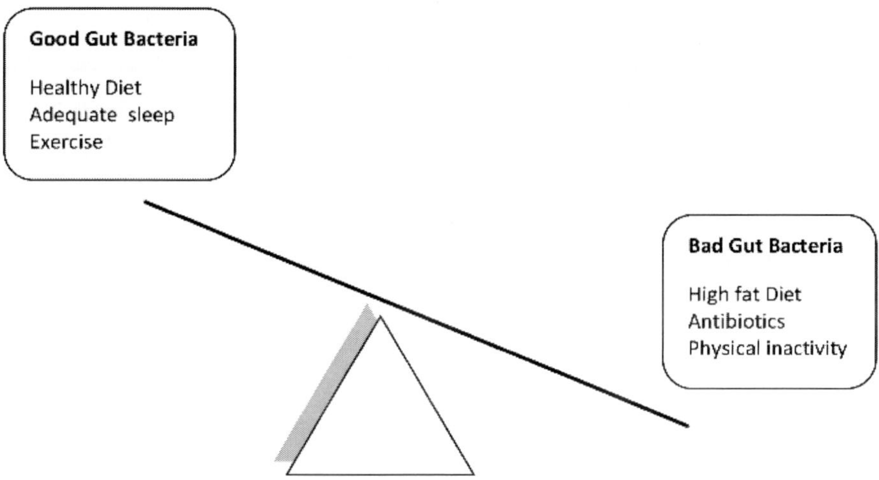

Figure 1. The delicate balance of our gut-flora.

Non-Modifiable Factors Affecting the Gut-Flora

Mode of Childbirth

The fetal intestines are usually believed to be devoid of bacteria. However, we now have recent scientific evidence that the oral cavity and meconium (the first stools of a new-born) has thriving communities of bacteria (Younge et al. 2019). The bacteria start to colonize our intestines even before our birth. During the passage of the fetus through the birth canal, the bacteria present in the mother's vagina gain access to the fetus through the oral cavity. You get your first bacteria from your mother. And her body even prepares for it as there is evidence that a particular type of

Lactobacillus dominates in a woman's vagina during pregnancy. Lactobacillus is the predominant bacteria present in curd. It digests lactose, which is the natural sugar present in milk. As milk is the only food for a new-born and the intestines are not fully developed, it is nature's way of ensuring the supply of nutrients to a new-born. So what happens to the babies born through cesarean section, which are continuously on the rise? The babies born through cesarean section have a different gut-flora composition as compared to those born normally. Scientific studies have shown that the babies born vaginally have gut bacteria, which resemble their mothers' vaginal bacterial communities. And if they are delivered by cesarean, their bacteria resemble those found on adult skin (Dominguez-Bello et al. 2010). Cesarean delivered babies have less diverse gut-flora and a low abundance of Bacteroides, Bifidobacteria, and Lactobacillus. They also take longer to acquire good intestinal bacteria, which help digest milk. However, these differences are most profound during the first three months of life and most disappear after six months of infant's life (Rutayisire et al. 2016). It is also observed that cesarean births are associated with a large number of diseases, particularly asthma and allergies, which are related to the immune system. Could gut-flora be the reason? The latest research indicates that, but more studies will probably give the definitive answer (Gallazo et al. 2020). We shall learn about the role gut-flora plays in the development of our immunity in chapter 2.

Genetics

The other non-modifiable factor affecting the composition of gut-flora is genetics. However, it plays a minor role compared to environmental factors. Studies conducted in twins have revealed that although the initial gut-flora is similar in twins, but as they age, it becomes different, indicating a strong impact of the environmental factors such as diet and exercise. Though a few types of bacteria are more influenced by genetics and considered heritable, the vast majority are not.

Ethnicity

The gut-flora markedly differs among people of different countries and ethnicities. The gut microbiome of tribal populations differs from that of the modern-civilization dwellers, being characterized by an abundance of bacteria that digest complex plant carbohydrates. The Hadza people's gut microbiome who are one of the last surviving hunter-gatherer tribes and inhabit the Eyasi valley in northern Tanzania is amongst the most diverse. Their gut-flora is shown to differ from that of most modern populations (Smits et al. 2017). Similarly, the gut-flora of Americans is different from that of Indians, and Indians' from that of Chinese and Japanese (Dhakan et al. 2019). Most of these differences are assumed to be due to different dietary habits, but there might be some as yet unknown factors too.

Aging

The composition of our gut-flora varies markedly with our age. As we know, the development of gut-flora starts right at birth. However, it is relatively simple at birth. It quickly grows to become a truly complex and diverse community that represents an adult microbiome. At the time of birth, we all have very similar gut-flora, the difference being only due to the cesarean delivery. Your gut-flora was similar to the baby lying in the adjacent cot. But there is a huge difference between us by the time we are grown-ups. Now, if all the people at your work-place decide to look at their gut-flora, the chances are that less than 10 percent of your gut-flora would be similar to any of your colleagues. Maybe a little bit more with the best-friend you share your lunch with. In a recent meticulously conducted study, researchers studied the gut-flora of infants at regular interval till school age and mapped the changes occurring with age. During early infancy, the gut-flora was markedly similar up to 3 months of age. It became diverse gradually till around seven months of age. After seven months, there was a rapid increase in microbiome diversity. They also found that among the major groups of gut bacteria, Bacteroidetes and

Proteobacteria were present as early as one month of age whereas Firmicutes were late arrivers. At school age, the gut-flora was diverse, with all the four major groups establishing their hold. At the age of around 12 weeks, the diet becomes the most crucial factor in the development of gut-flora (Galazzo et al. 2020). Fast forward a few decades, you are happily retired. Today is the reunion day with your old-mates. And someone throws this quirky idea of looking at the gut microbiome again. Now, your gut-flora would be more similar to that of your old-mates. As we get old, our gut-flora again becomes less diverse. At both the beginning and the end of our life-time, gut bacteria are less diverse. Proteobacteria also increase in the gut-flora of the elderly (Xu et al. 2019).

Modifiable Factors Affecting the Gut-Flora

We have shed some light on the important non-modifiable factors. But you would be wiser after reading about the factors that we can alter or modify. These are the ones you can work on to achieve a healthy gut-flora. You would be pleased to know that the modifiable factors have a larger impact on our gut-flora than the non-modifiable ones. The modifiable factors which affect our gut-flora are:

Breast Feeding

To the hundreds of benefits of breast milk you already know, let us add one more. Breast-fed babies have healthier gut-flora. They have a higher number of Bifidobacterium in their intestines (Bezirtzoglou et al. 2011). Bifidobacteria consume special sugars found in breast milk, called human milk oligosaccharides, which babies cannot otherwise digest. As a result, these bacteria are present in breast-fed infants' guts but not of formula-fed ones. Bifidobacterium is a significant resident of our intestines, as we shall learn later. Breast milk is a 'living' fluid. It is full of bacteria and also contains antibodies and hormones. Not only infants receive the bacteria from the milk but also from the areola while sucking. If you are a soon-to-be mother, you can lay a strong foundation for your child's gut-flora and her health by merely feeding her your own milk.

Gut-Flora and Asthma/Allergies in Children

You will read about the changes in gut-flora concerning illnesses in section 2. But let us discuss asthma and allergic disorders in children as they are often associated with cesarean delivery and lack of breastfeeding. In the study by Gallazzo et al. the researchers found the bacterial species that were differentially abundant among infants who did or did not develop the allergic disease (AD). Lachnobacterium and Faecalibacterium were the two bacteria, decreased throughout infancy among children who developed AD. Besides, Lachnospira and Dialister were reduced among children who developed asthma. The fact that these bacteria were reduced throughout infancy suggests their role in the protection against allergic disease. Similar findings were also observed in the Canadian Healthy Infant Longitudinal Development (CHILD) Study, which revealed Lachnospira and Faecalibacterium to be decreased among children at the risk of allergic wheeze at the age of 1 year (Subbarao et al. 2015).

Are these findings a mere chance? Or is there any biological basis behind them? Acetate, a short-chain fatty acid, is one of the fermentation products of Lachnospira and Lachnobacterium. Both these bacteria are reduced in the gut-flora of children with allergies or asthma. They produce acetate in our intestines from the undigested carbohydrates. Studies conducted in mice have shown that acetate-feeding leads to a marked suppression of allergic airway disease in a mouse model for human asthma (Thorburn et al. 2015). Acetate acts at the genetic level to reduce the inflammation in the airways. Faecalibacterium also has anti-inflammatory effects, which occur through the production of butyrate and a microbial anti-inflammatory molecule.

Diet

Out of the many factors that influence our gut-flora composition, the diet certainly stands out as one of the most significant players. And yes, it can be modified! Our diet is the link through which we can tweak our gut-flora and enjoy excellent health! The relationship between gut microbiome and diet is not one-sided but bilateral. While what we eat impacts the composition of our gut-flora, HOW what we eat is utilized by the body depends on the design of gut-flora!! It is a give and take relationship-mutualism. Our gut bacteria aren't altruistic after all!

To understand the impact of diet on gut-flora composition, let us first split the diet into individual components and understand their impact and subsequently put the pieces together to build the complete picture. Our diet contains macronutrients, which are carbohydrates, proteins, and fats. As the name suggests, they form a major portion of what we consume. The other portion is micronutrients, which include various vitamins, minerals, antioxidants etc. which are required in much smaller amounts for the functioning of our body. Let us look at how they impact the gut-flora.

Carbohydrates

Out of all the macronutrients, carbohydrates have the most significant impact on gut-flora composition. Although 'the simple is the best' rule applies to most domains of our lives, the relationship between the carbohydrates and their impact on gut bacteria stands out as an exception. Moszak et al. have extensively described this relationship. Simple carbohydrates such as fructose and sucrose harm the gut-flora, while the complex carbohydrates remodel it to our benefit!

Simple carbohydrates in diet= Adverse impact on gut-flora composition.

Complex carbohydrates in diet = Favorable impact on gut-flora composition.

Are any complex carbohydrates in particular? Yes. A specific subtype called 'Microbiota-accessible carbohydrates' (MACs) have a significantly beneficial impact on the gut microbiome composition. These MACs are found in fruits, vegetables, whole grains, and legumes.

Carbohydrates are an integral part of the human diet. When we ingest carbohydrates, their digestion begins in the mouth through the action of the enzyme 'salivary amylase'. The carbohydrates are further digested in the small intestine. However, the case with MACs is different. These are resistant to digestion by our enzymes and instead act as the substrates for metabolism by the gut bacteria. These are also termed prebiotics. The gut bacteria ferment and metabolize these into compounds such as Short Chain Fatty Acids (SCFA), which have health benefits for us.

Now that we understand what MACs are, let us explore some fun facts about them!

'Having the same food is not the same as having identical amounts of MACs for two individuals!!'

Identical food≠ Identical MACs

Let us explain with two interesting examples.

Example 1: For most Japanese individuals, porphyran, an algal complex carbohydrate, acts as a MAC as their gut bacteria have the capacity to metabolize it. The same molecule is NOT a MAC for the majority of North American and European population as their gut-flora lack the genes required to metabolize porphyrin.

Example 2: Lactose, the sugar found in milk, is metabolized by the enzyme 'lactase' in the small intestine in case of most of us and isn't available as a substrate to the gut-flora. However, in case of those who are lactose intolerant, lactose escapes digestion in the small intestine and acts as a MAC!!

Now we know what happens when we feed our tiny friends with MACs; they get happy and bless us with a healthy immune system and keep lifestyle diseases at bay. But what happens when we don't feed them? Well, they turn vengeful!! Dangerous!!! When the diet lacks MACs, the microbes start feeding on the other readily accessible carbohydrate source: Us! Yes. This is true!! Our intestines are lined by a slimy mucus coat that is rich in carbohydrates. When the gut bacteria start feeding on that mucus coat when unable to get hold of MACs from our diet, the immune system gets alerted- Beware! Beware!! Beware!!! A microbe is penetrating the intestinal wall! If the diet continues to lack MACs over an extended period of time, the immune system becomes 'edgy' and results in a host of health problems.

Proteins

Proteins, just like carbs and fats are essential component of a healthy human diet. About 10-20% of our daily calorie requirement is met through the intake of proteins. Proteins have many vital functions in our body. The gut-flora is capable of metabolizing the proteins and generates SCFAs, amines, hydrogen sulfide, phenols, and CO2, etc.

The impact of protein intake on gut-flora depends upon the concentration of the protein, its source, and also the composition of amino acids. Scientists have investigated the impact of different dietary protein sources on gut-flora composition. Casein, along with the other dairy-based proteins, was found to have the most beneficial impact on the composition of the gut microbiome, which was speculated to be because of the high branched-chain amino acid content (Madsen et al. 2017). The intake of these proteins was also associated with the prevention of obesity. Also, the intake of seafood versus meat is associated with different compositions of the gut microbiome, with the former correlating with a healthy gut microbiome while the latter being associated with obesogenic and disease-causing bacteria.

We all know what happens when we do not consume proteins in adequate amounts, but have you ever wondered what happens when we consume proteins in excess? It is a common misconception that excess of protein doesn't lead to any adverse health consequences, and there is nothing like 'protein excess' in the diet. However, the rule 'excess of everything is bad' applies to protein intake as well. If a large amount of protein remains accessible to the gut bacteria for an extended period of time, there is an undesirable alteration in the composition of gut microbiome, increasing the number of disease-causing microorganisms. A high protein diet is associated with an increase in the number of pathogenic organisms that cause metabolic illness. In a randomized pilot study conducted by Perez et al. 2018, cross-country runners were administered a diet supplemented with whey isolate and beef hydrolysate for ten weeks. This was found to result in an increase in the concentration of the *Bacteroidetes* and a decrease in the concentration of beneficial bacteria such as *Roseburia, Blautia*, and *Bifidobacterium longum*. Hence, it should be remembered that too much of a good thing is bad, too, and protein intake is no exception to the rule!

Fats

As per the dietary recommendations by the eminent bodies involved in designing nutritional recommendations globally, fats or dietary lipids should ideally fuel the body with approximately 30% of the daily calorie requirements. Dietary lipids include three types of fatty acids and their derivatives:

- Saturated fatty acids (SFA)
- Monounsaturated fatty acids (MUFA)
- Polyunsaturated (PUFA) fatty acids and their derivatives, including phospholipids and cholesterol.

Fats are the most calorie-dense of all three macronutrients. Although most of us view fat as the culprit behind making us fat, it has many vital functions in our body, but as is the case with other macronutrients, its excessive consumption may lead to adverse health effects.

The impact of fat consumption on gut-flora depends on both the quantity as well as the composition of the lipids. High-fat diet alters the gut-flora. It has been shown to lead to reduced diversity as well as an increase in the phyla Firmicutes and Proteobacteria in mice (Hildebrandt et al. 2009). In an elegant study conducted in China, researchers fed young, healthy individuals an iso-caloric diet with different fat quantity. They randomly assigned these individuals to receive a low fat (20% of energy through fats), moderate fat (30% of energy), and high fat (40% of energy) diet. They found that the high-fat diet group individuals had reduced diversity of bacteria in their gut-flora. All the subjects received an equal number of calories. It confirmed the effect of fat on the gut-flora is independent of the calories consumed. Besides, the concentration of total short-chain fatty acids was reduced in the high-fat diet group compared to the other groups (Wan et al. 2019). The composition of fats also impacts gut-flora. The ratio of SFA: MUFA: PUFA plays a crucial role in determining the gut-flora composition. A diet with which is high in MUFA and PUFA results in a higher Bacteroidetes: Firmicutes ratio and lactic-acid bacteria (Costantini et al. 2017).

Additional Factors

Apart from carbohydrates, proteins, and fats, there are other components of the diet that collectively have a significant impact on gut-flora composition. These include tea, coffee, alcohol, carbonated beverages, salt, vitamin supplements, polyphenols, and fermented food rich in probiotics. Let us talk about the two important components, polyphenols, and probiotics.

Polyphenols

Dietary polyphenols are the substance found in tea, fruits, vegetables, and cocoa, which are known to possess antioxidant properties. What are these antioxidants?? Well, antioxidants are those substances that have the potential to slow down/prevent the damage to the cells and tissues in our bodies that results from the action of free radicals generated as a consequence of environmental and other stressors. Polyphenols include compounds with a diverse chemical composition such as flavonols, flavones, phenolic acids, and anthocyanin, to name a few. The dietary intake of polyphenols favors the growth of Bifidobacterium and Lactobacillus, having a beneficial impact on gut-flora composition. Additionally, it also reduces the number of pathogenic bacteria such as *Clostridium perfringens* and *Clostridium histolyticum* (Ma and Chen 2020).

Probiotics

The World Health Organization defines probiotics as live microorganisms which when administered in adequate amounts confer a health benefit on the host. Probiotics exert their health-enhancing action by restoring the healthy gut-flora or altering its composition. The concept of probiotics was introduced by the Nobel laureate Élie Metchnikoff, who highlighted the link between yogurt-consumption and increased longevity in Bulgarian peasants.

These probiotics are abundant in fermented foods and are also marketed as health supplements. Do probiotics health supplements work? We will explore the answer in subsequent sections, but as of now, we will focus on the food sources of probiotics. Fermented foods are one of the best sources of probiotics in the human diet. Humankind has been fermenting food for centuries. This is how the process of fermentation works- the sugar and starch present in the food are metabolized by the bacteria giving rise to lactic acid. Fermented foods improve the gut-flora by simply providing healthy bacteria to our intestines. Different sources of probiotics are popular in different parts of the world. Talking about India in particular, the curd is one of the most extensively consumed probiotics that is traditionally set at home and is also integral to many of the traditional religious rituals too! Some of the common examples of probiotics include – Curd, Yogurt, Pickles, Miso, Sauerkraut, and Kimchi.

We just learned about the impact of different dietary components on gut-flora structure and function- carbohydrates, proteins, fats, polyphenols, and probiotics. This has set the stage for us to understand what our diet should look like and the impact of fad diets on gut bacteria. Utilizing this knowledge that we have just acquired, we will learn about these topics in section 3.

Let us now move from the dietary components and look at how the diet as a whole impacts gut-flora. The Western diet is a dietary pattern that includes high intakes of red meat, pre-packaged processed foods, fried foods, refined grains, corn (and high-fructose corn syrup), and low intakes of fruits, vegetables, whole grains, fish, nuts, and seeds. It is rich in saturated fats and refined, processed sugars and poor in unsaturated fats and fiber. Most of the experts believe that the Western diet is the reason for the increase in lifestyle diseases. In a Korean study, researchers compared the gut-flora in individuals who consumed a western diet for four weeks and then crossed over to the traditional Korean diet rich in fish, fiber, and fermented foods. They found that after the four weeks of the western diet, the gut bacteria diversity decreased while it increased after the traditional Korean diet (Shin et al. 2019). Firmicutes were increased, and Bacteroidetes were reduced in gut-flora after the Korean diet. It is interesting to know that a particular bacterium Weissella was increased by the Korean diet. Weissella is a probiotic bacterium that is crucial in the fermentation of Kimchi, a traditional fermented vegetable dish in Korea (Jung et al. 2011). The fact that the western diet could alter the gut-flora in Korean individuals highlights that diet has a much greater impact on gut bacteria than ethnicity.

The Mediterranean diet is considered a particularly healthy diet as it has been shown to reduce the incidence of coronary heart disease over a long period of 25 years! (Kromhout et al. 1995) The Mediterranean diet was first defined by Ancel Keys as being low in saturated fat and high in vegetable oils, particularly olive oil, and is consumed in Mediterranean countries. It varies by country and region, so it has a range of definitions (Davis et al. 2015). But in general, it is high in vegetables, fruits, legumes, nuts, beans, cereals, fish, and unsaturated fats such as olive oil with a low intake of red meat, saturated fat, and dairy foods. Let us see if the Mediterranean diet has any special impact on gut-flora. De Filippis et al. showed that adherence to the Mediterranean diet increases the Firmicutes in the gut-flora and also leads to increased production of short-chain fatty acids (De Filippis et al. 2016). Another group of scientists showed that it also increases the Bifidobacterium, our valuable friend in the gut-flora (Mitsou et al. 2017).

Exercise

Most of us are aware of the beneficial impact of exercise on the cardiovascular system, endurance, strength as well as muscle and bone health. However, very few know about its impact on gut-flora. Well, yes, it might be difficult to digest, but exercise does alter the gut-flora. In fact, exercise has an independent effect on the composition of gut bacteria meaning that exercise improves gut-flora irrespective of a person's diet and Body Mass Index (BMI). The exact mechanism of how it works is not yet understood but let us look at some scientific studies that provide insight into this.

It was first noted in 2008 that exercise might independently alter the gut microbiota through studies in laboratory mice. A group of scientists from Japan compared the gut-flora between sedentary (non-exercised) and exercised mice (Matsumoto et al. 2008). Wondering how one can make the mice exercise? The exercised group was placed in cages with wheels with a design that forced mice to use them. The microbiota of the exercised mice exhibited increased production of short-chain fatty acid (SCFA), butyrate. Butyrate production is possibly the most critical function performed by the gut bacteria. We will read about butyrate and its importance in more detail later in this chapter.

Illustration 1.

The beneficial effects of exercise on gut-flora have been evaluated in human studies as well. According to the latest research, exercise training beneficially modifies gut-flora composition. A study was conducted on international rugby players in Ireland, wherein their gut-flora was compared with that of healthy adults from the Irish population. It demonstrated that the rugby players who train and exercise religiously had increased diversity in their gut-flora. In particular, there was a greater diversity among the Firmicutes phylum that help to maintain a healthy gut environment (Clarke et al. 2014). In another extensive study, Estaki et al. (2016) analyzed the fecal microbiota of 39 individuals with different fitness levels and comparable diets. Peak oxygen uptake (VO_2 max, also known as maximal oxygen consumption, maximal oxygen uptake, or maximal aerobic capacity) is a gold standard indicator of cardio-respiratory fitness and endurance. It is the maximum rate of oxygen consumption measured during incremental exercise; that is, the exercise of gradually increasing intensity. A high VO_2 max means that the person can utilize an increased amount of oxygen. The higher the VO_2 max, the more oxygen your body can use – and the better your cardio-respiratory or aerobic fitness. This study indicated that regardless of the diet, cardio-respiratory fitness was correlated with increased gut microbial diversity, thus establishing the independent role of exercise in altering gut-flora in humans. Furthermore, the individuals who exercised were found to possess a microbiota enriched in butyrate-producing groups of bacteria, such as Clostridiales and Lachnospiraceae, resulting in increased butyrate production, an indicator of good gut health.

Now, we understand that exercise independently affects our gut-flora. But how much duration of exercise is required to bring about these changes? The beneficial changes in the gut-flora can be observed as early as six weeks after beginning the exercise (Allen et al. 2018). These changes manifest as a decrease in the density of inflammation-causing bacteria (e.g., Proteobacteria), and an increase in the density of bacteria linked with enhanced metabolism (e.g., Akkermansia). Sedentary individuals who initiate moderate-intensity exercise develop a greater diversity of gut bacteria, especially phylum Firmicutes. *Exercise* enriches bacterial diversity; improves the Firmicutes-Bacteroidetes ratio, which may potentially contribute to weight loss and alleviate obesity-associated diseases. It increases the number of bacteria that enhance mucosal immunity and intestinal barrier functions, reducing obesity and metabolic diseases (we will read about it in more detail later in the next chapter). Exercise also increases the density of bacteria capable of producing compounds that protect against gastrointestinal disorders and colon cancer (such as SCFAs). Therefore, we can use exercise as a medium to achieve and maintain a healthy gut-flora in as little as six weeks! Isn't that amazing?

SLEEP

Taking about gut health, exercise isn't the only surprise 'good guy'. Sleep affects the gut microbiota as well! The relationship between the two is not only crucial but mutual. Lack of sleep alters the gut-flora, that in-turn affects our sleep and mood. As little as two days of sleep deprivation can alter the gut-flora, upsetting the Firmicutes-Bacteroidetes ratio, as observed in a Scandinavian study. This led to decreased insulin sensitivity and an increased propensity to develop diabetes in these individuals (Benedict et al. 2016). It is not just the lack of sleep that affects the gut-flora. Getting fragmented sleep or sleeping at unnatural times can also alter the gut-flora. This is particularly a problem for rotating shift workers and frequent flyers, whose microbial health often suffers from repeated sleep cycle disturbances, even when they can manage sufficient numbers of sleep hours. A study conducted in mice revealed that just four weeks of altered sleep cycle reduces the diversity of gut-flora and alters its balance (Deaver et al. 2018).

ANTIBIOTICS

Antibiotics are essential, life-saving drugs that are taken to cure diseases resulting from bacterial infections. Most of us have consumed antibiotics to treat an infection. But apart from killing the disease-causing bacteria, the antibiotics also kill the friendly bacteria present in our gut-flora. Antibiotics cannot differentiate between the two. The consumption of antibiotics decreases the number, as well as the diversity of gut bacteria. The antibiotic consumption alters the gut bacterial composition within 3–4 days of starting the drug. However, the good news is that after one week of discontinuing antibiotics, gut-flora begins to recover and achieves an almost similar pre-antibiotic composition in a month's time. But it takes more than six months for gut-flora to fully recover after a course of antibiotics.

One of the most important functions of gut bacteria is to inhibit the growth of pathogenic bacteria by competing with them for nutrients. Since they have a hold on our intestines, they don't let the other bacteria to establish themselves. Loss of gut-flora due to the use of antibiotics provides an opportunity for disease-causing bacteria to grow, leading to gastrointestinal infections. One such example of opportunistic pathogenic bacteria is *Clostridium difficile*, which typically causes infection in the setting of antibiotic use. Due to the increased use of antibiotics, its incidence is on the rise across the globe.

The take-home message- Don't fear antibiotics but don't abuse them either. Use them judiciously AS PRESCRIBED BY YOUR PHYSICIAN. Self-medication is dangerous and can do you more harm than good.

OTHER MEDICATIONS

Apart from antibiotics, a lot of other medicines also alter our gut-flora. In an extensive study investigating the impact of non-antibiotic medicines on gut-flora, out of 835 non- antibiotic medicines, around 200 restrained the growth of at least one type of bacteria in the human gut and around 5 % of medicines affected at least ten different types of bacteria (Maier et al. 2018). Some commonly used medicines which alter gut-flora are anti-inflammatory drugs (commonly known as pain-killers), antipsychotics, cancer drugs, proton-pump inhibitors, and Metformin (used in diabetes). These medicines decrease the diversity of the gut-flora.

The anti-inflammatory drugs and proton-pump inhibitors (PPI) are the most commonly used prescription and over-the-counter drugs. Each of us has popped these into our systems without considering and bothering about the prescriptions. We feel happy about the instant relief while forgetting to consider the side effects. One of the not-so-great effects of both these classes of medicines is an alteration of our gut-flora.

PPI are drugs that inhibit acid secretion in our stomach. It is widely consumed for the management of acidity, bloating, and heart-burn. In a landmark study wherein the stool samples of the participants taking PPI were analyzed, a decrease in the diversity of the gut bacteria was observed. PPI consumption in participants altered 20 % of the bacterial species present in the intestines (Imhann et al. 2016). Furthermore, PPI consumption increased the presence of oral bacteria in the intestines. As the oral bacteria pass into the stomach along with the ingested food, they are killed by the acid present in the stomach. However, upon PPI consumption, the acid secretion is suppressed, and these bacteria reach the intestine. While the presence of the oral bacteria in the intestines is not a cause of concern in healthy individuals, in the case of people with weakened immune systems suffering from chronic illnesses such as diabetes and liver cirrhosis, these oral bacteria can cause infections. The PPIs are highly effective medicines, but, just like antibiotics, should not be abused and taken only as per physician's advice.

Another very important commonly used class of drugs affecting gut-flora are non-steroidal anti-inflammatory drugs (NSAIDs), commonly called pain-killers. They also reduce bacterial diversity and increase the representation of potential enteric pathogens in the gut-flora (Rogers and Aronoff 2016). The effect of medications on gut-flora may be beneficial too. Metformin, the most commonly used medicine for the treatment of diabetes, alters gut-flora leading to the increased short-chain fatty acids in our intestines, which may even contribute towards its beneficial effect (Wu et al. 2017).

After looking at the major factors that affect the gut-flora composition, we can conclude that the most important determinant of gut health is our lifestyle. Be it diet, exercise, sleep, or the medicines, we can employ them judiciously for the best outcomes. In most people, the aches and pains or acidity and heart-burn are due to a poor lifestyle, which we have the power to alter. Anything that we put into our mouth affects the bacteria present in our intestines. But why should one put in efforts to improve the gut-flora? Well, your question is understandable. After all, it sounds sensible to exercise to lose weight, look slim, build muscles, but exercising to grow some unpronounceable and weird bacteria in intestines seems over-the-top, doesn't it? Well, your perception is about to change. Shortly, you will learn about all the work these bacteria do for us. Then you wouldn't mind working hard to grow them at all!

Chapter 2

GUT BACTERIA: WHAT THEY DO FOR US?

When in school, our teachers taught us that some bacteria reside inside our intestines where they make vitamin B and vitamin K. That was it. We, the authors, graduated and post-graduated in the medical and the life sciences, but our knowledge about the gut bacteria was still limited. We did not know that the synthesis of these vitamins by gut-flora is not even the tip of the iceberg in the vast functions these bacteria perform. Bacteria do way too much more for us than simply make a few vitamins. Well, neither our esteemed teachers nor we were at fault for our ignorance. Nobody knew much about the gut bacteria. The bacteria can be present in human intestines as a part of normal flora was first discovered in mid-1880s when Austrian pediatrician Theodor Escherich identified *Escherichia coli* in the fecal samples. However, Prof. Escherich himself concluded that these bacteria were just bystanders and may not play any role in our digestive process. After that, the knowledge about gut bacteria lay dormant for the next 120 years. In 2008 when the *Human Microbiome Project* was launched, we started to gain an exciting and tremendous amount of knowledge about the bacteria that reside inside or on our body. Although scientists have been studying the role of bacteria in various human diseases, the ways to study them were limited. To study them, bacteria needed to be grown in artificial media called cultures. But there was a major limitation. We cannot grow all the bacteria in artificial cultures. A major breakthrough occurred with the launch of the *Human Genome Project,* which was completed in 2003 to identify all the genes we possess. It led to the development of technologies that enabled to study bacteria without culturing them. Another and perhaps the most critical information which Human Genome Project told us was that human beings carried only 30000 genes, which are almost the same as in Drosophila, a fruit fly. It was estimated at the start of the human genome project that we must be carrying around 100 000 different genes to carry out all the complex functions which the human body can perform. This finding led to the startling revelation that perhaps a large number of functions in the human body are actually being performed by the micro-organisms living inside us, and that information is embedded in their genes rather than ours!

Thus, only recently have we started to gain insight into the world living inside us and appreciate its contribution to our own well-being. The knowledge about the gut-flora is new and is not well known to most of us. Let us read about the various functions gut -flora performs for us.

Energy Absorption

The food we eat provides us with a variety of nutrients and calories. A calorie is simply a measure of energy. We may loathe calories and look for 'low-calorie' alternatives to everything we eat, but the fact is we need the energy to perform each of our tasks, even to breathe. Experiments in mice have revealed that we need gut bacteria for optimum absorption of calories from our food. Scientists create germ-free mice by raising them in a sterile bubble so that they are not exposed to any microbes. These mice don't have any microbes in their body, including the gut-flora. In a study, the germ-free mice were compared with the conventionally raised mice for their weight and feeding patterns. At a similar food intake, the weight of the germ-free mice was found to be significantly less than that of the normal mice (Bäckhed et al. 2004). Extrapolating it to human beings, in the absence of the gut-flora, we will have to consume a much higher amount of food to maintain our usual bodyweight. However, it does not mean if we get rid of our gut-flora, we can drop weight and treat obesity. On the contrary, an alteration in the healthy gut-flora results in over-weight and obesity. We shall learn about it in subsequent chapters.

How do the gut microbes release calories from food?

As we know, carbohydrates are the chief source of energy for the human body. However, a large proportion of food that we consume includes polysaccharides, the complex carbohydrates that cannot be digested[2] because we lack the enzymes required to break them down into simple carbohydrates. The gut-bacteria produce specialized enzymes which break-down these polysaccharides into simpler forms, which can be absorbed by our body. One such efficient friend of ours is *Bacteroides thetaiotaomicron*. Nearly 20% of its DNA is dedicated to the production of enzymes that break down complex carbohydrates (Martens et al. 2008). The human body does not produce these enzymes. Without these bacteria, all the calories present in the complex carbohydrates would be lost. It is estimated that these complex carbohydrates contribute 10%-30% of the total energy we get from our food. Do you know? The bacteria present on the floors and the machinery of sugar mills are similar to those found in our gut. They, too, have an abundance of Firmicutes and Bacteroidetes. Well, indeed, the break-down of the complex sugars is one of the most important functions of the gut-flora, after all.

Maintaining the Integrity of the Gut Barrier

Our intestines are like hollow tubes, their insides lined by a layer of cells called the epithelium. The individual cells comprising the epithelium are just like the stones on a pavement held together by various micro-

---

[2] Digestion, in the most simplistic terms, is the process of breaking down large, complex compounds present in food into the small, simple compounds which can be absorbed in our intestines.

structures acting as cement. Although these cells adhere closely to each other, there is space between them. The bacteria can pass through these spaces and reach the blood vessels or the lymphatic system below the epithelium. This property of intestines to allow certain molecules/particles/ bacteria to pass through the gut epithelium is termed as the *gut permeability*. You must know that the gut permeability isn't a fixed entity. It changes in response to different stimuli. When the gut permeability increases, many undesirable substances can pass through, resulting in disease and sickness. The gut-flora, the unsung hero, protects us by maintaining the gut permeability by keeping the gut barrier robust!

How gut-flora keeps the gut barrier intact?

The gut barrier is comprised of the epithelial cells, along with various micro-structures which bind them, and a layer of mucous. The gut bacteria help in maintaining the gut barrier function in several ways. The most important of these is the production of short-chain fatty acid, Butyrate. Butyrate is the chief source of energy for the intestinal epithelial cells and is required to keep them healthy. Healthy epithelial cells keep the gut barrier intact. Butyrate is solely made by the gut-flora. It acts at the genetic level to promote the production of various molecules that act as cement to bind the cells together (Peng et al. 2009) and increase mucous production (Shimotoyodome et al. 2000). Apart from butyrate production, gut bacteria themselves interact with the epithelial cells through specialized molecules and promote their proliferation, which contributes to the healing of the gut barrier after an injury. (Thomas et al. 2018). Almost all the lifestyle diseases we encounter are due to an increase in inflammation in our bodies. One of the ways the inflammatory molecules gain entry into our body is through our gut. An inefficient gut barrier results in an increased flow of such substances from the intestinal cavity into the blood vessels or the lymphatic ducts. The gut bacteria help in reducing the entry of these inflammatory molecules by keeping the gut barrier intact.

DEVELOPMENT AND MAINTENANCE OF IMMUNITY

'Immunity' has become buzz word, especially due to covid-19. If you have good immunity, can you fight covid-19 better? Whether the infection with covid-19 leads to everlasting immunity? Will the vaccines provide immunity against SARS CoV-2? Everyone has questions about immunity. But what is immunity?

Immunity is the ability of our body to resist infections. Our body has different organs where specialized cells reside which fight against the microbes; this anti-microbial machinery of ours is termed as the immune system. It is interesting to note that gut bacteria play an important role in determining the state of the immune system. Gut bacteria are selfless, but before you assume that they sacrifice their life for our well -being, let us tell you that our immune system doesn't kill them. Why does the immune system, whose primary function is to protect us from infections, doesn't fight against the gut bacteria? Confused? Read on to know more.

Our immune system works in a two-step process. To protect us from infections, first, it has to identify the microbes and then eliminate them. The immune system has a remarkable capability to distinguish self-antigens from non-self antigens, identifying the microbes. What are antigens? Any molecule that can activate the immune system is termed an *antigen.* It can be a protein, or a polysaccharide (complex carbohydrate), or a combination of the two. Self-antigens are our own antigens that the immune system recognizes as harmless, and hence, it doesn't attack them under normal conditions. The immune system's inability to identify self-antigens results in *autoimmune diseases,* in which the immune system starts attacking the self-antigens as if they are alien and dangerous.

The non-self antigens are those that come from the external environment. The various pathogenic organisms fall into this category. These need to be identified and eliminated by the immune system to prevent infections. The story is not that simple, though. We, the humans, don't live in complete isolation. We are just a part of a complex ecosystem wherein we continuously interact with the environment. Can you imagine that we encounter a vast number of external non-self antigens every day?

We inhale many other molecules apart from oxygen, such as pollen-grains, dust particles, and pollutants. In addition to the airways, there is another even bigger portal of entry for non-self antigens into the human body. Yes! That is our gut. We eat food every day, which has molecules that are not intrinsic to our human body. However, our immune system does not mount a response to eliminate them. If the immune system were to attack them, we would be in a state of continuous war against these antigens and spend a major chunk of our energies doing just that. There would be no growth, and it would be impossible to survive. The human body already knows that no growth occurs during a war; if only human mind would have known that too! Similarly, a huge number of bacteria, which are clearly non-self, also reside in our intestines as a component of gut-flora without mounting an immune response, and this is how it happens.

The immune system broadly comprises cells that can directly kill the pathogens, a special type of molecule which helps in killing the pathogens called *antibodies,* and the cells which help in the synthesis of antibodies. We know that these cells reside in specialized tissues such as lymph nodes, spleen, and thymus, which together form the structural framework of our immune system. But a large number of immune cells are also present in our intestines. And in-fact, the gut is the largest immune organ of the human body. So what is this specialty of the gut immune system so that even when it recognizes a huge number of non-self antigens, it does not respond and try to eliminate them? The cells and antibodies comprising the gut immune system respond differently from the immune cells present in other lymphoid organs. So, when an external antigen is present in the gut lumen, it activates a unique immune cell *Tregs* suppressing the further activation of the immune cascade. Some of the friendly bacteria present in our gut-flora, such as *Lactobacillus reuteri* and *Lactobacillus casei* can lead to increased production of Tregs and thus help in proper functioning of the gut immune system (Smits et al. 2005). *Bifidobacterium infantis* has also been shown to increase the Tregs in healthy human volunteers (Konieczna et al. 2012).

But, the gut immune system also needs to identify the harmful pathogenic organisms and respond to eliminate them. The gut bacteria play an essential role in the development of the gut immune system by providing antigenic stimulation to the immune cells during infancy and early childhood (Dzidic et al. 2018). The gut-flora trains our immune system. This training of the immune cells by gut bacteria is essential as it has been shown that germ-free mice are more susceptible to infection by intestinal pathogens than normal mice. As we are born, we are protected from infections by the antibodies transferred to us from our mothers. Our own immune system is naive and not fully developed. The bacteria present in our gut educate the immune system. When germ-free mice are provided with bacterial flora during experiments, they develop more diverse immune cells. This education starts right after birth when gut-flora begins to grow. Besides, gut bacteria also teach our immune system that not all external substances are harmful, and in fact, many of them are essential for our survival.

Illustration 2.

PROTECTION FROM THE DISEASE-CAUSING BACTERIA

The friendly gut bacteria strengthen the immune system and actively block the growth of harmful pathogenic bacteria. Both the friend and foe bacteria require nutrients to grow and multiply. The friendly bacteria compete with the pathogens for the nutrients as well as space inside the gut. Reduction in good gut-flora due to poor diet, disease, or antibiotic treatment allows the gut to be available to disease-causing bacteria. As already mentioned in chapter 1, one such infection is *Clostridium difficile,* which occurs when a person receives antibiotics. Besides, the friendly gut bacteria also synthesize some chemicals that can kill harmful bacteria.

BRAIN-GUT AXIS

"See, I told you I had a gut feeling we would get this question for math's paper, and look, it did!" How often have you told something like this to your friends? We have all had 'gut feeling' for matters that are trivial as well as significant. But did you know there is A REAL connection between your brain and the gut? Yes, your gut does speak to your brain; it does! You better start listening to it. Do you still think we are kidding? Well, just remember the times when you are hungry. Simply at the sight of food, don't you feel those butterflies in your stomach? This is a two-way connection, though. The gut and the brain both affect each other. Many of us feel so angry when hungry that it has led to the coinage of a new word, "hangry", a combination of hungry + angry.

There are three ways by which interaction occurs between our gut and the brain. The gut-flora has an impact on all three of them.

1. Neural (related to the nervous system): The Vagus nerve originates in our brain, and its neurons (individual cells) are present throughout the digestive system. These neurons form synaptic connections with the neurons of the enteric nervous system (ENS).

ENS is the collection of neurons that are present in the wall of our digestive system. The nerve endings of these enteric neurons are continuously exposed to the gut bacteria. The signals from these neurons can alter the gut motility and secretion of gut hormones.

2. Immune pathway: As mentioned in the previous section, gut-flora interacts with the immune cells present in the intestines. These immune cells secrete various molecules, called cytokines and prostaglandin E2, which travel to the brain via the bloodstream and affect brain function.
3. Endocrine system: We know that hormones are the substances present in our bloodstream, affecting a wide range of our body functions. The cells which produce and secrete hormones are termed *endocrine* cells. Our intestines have more than 20 types of endocrine cells. That makes the intestines the largest endocrine organ in the human body. These endocrine cells lie in the close vicinity of gut-flora, which can drastically affect them. One such significant interaction is between Butyrate, produced by gut bacteria, and the endocrine cell producing Peptide YY. Butyrate stimulates these cells to increase the production of Peptide YY. This peptide travels to the brain through bloodstream tell the brain that you are full and need to stop eating. Never underestimate the power of common bacteria! You need bacteria to say to you when to stop eating.

Through the endocrine system, gut-flora even impacts the secretion of other hormones such as serotonin, the HAPPY hormone. The secretion of serotonin in the brain gives us the feeling of happiness and elevates mood. God! The gut bacteria control not only our satiety but our emotions too! A perfect balance in the gut can lead to a perfect balance in the brain, and peace and harmony in life. Happy gut=Happy life!

Isn't it humbling to know that the tiniest of creatures residing inside us perform so much required for our very survival? Gut bacteria impact our physical as well as mental and emotional well-being. They protect us from infections and lifestyle diseases such as diabetes, obesity, and high cholesterol. But we, the 'bacteriophobic' humans, are continually trying to get rid of them without realizing that neither it is beneficial nor possible. There are only a handful of bacteria that are harmful to a healthy human as opposed to the vast majority, which are valuable and necessary for our survival.

Illustration 3.

Short-Chain Fatty Acids

Short-chain fatty acids (SCFA) are an essential class of compounds. Fats, along with carbohydrates and proteins, are the primary macronutrients we need for our nourishment. Fats are the most energy-dense as they provide 9 kcal per gram compared to carbohydrates and proteins, both of which provide 4 kcal/g of energy.

The fats are either saturated fats or unsaturated fats. Saturated fats are solid at room temperature. They include all the animal fats like butter, ghee, the fat present in the red meat, and coconut oil. The unsaturated fats are liquid at room temperature. They include all the fats of plant origin except coconut oil and fish oils. There are a few fats which our body cannot make on its own. They are termed essential fatty acids. All the essential fatty acids are unsaturated fats.

While saturated fats receive constant criticism from doctors, nutritionists, and health practitioners for being unhealthy, there is a class of saturated fats that are both beneficial and essential for us. These are short-chain fatty acids or SCFAs.

The fats are compounds made of carbon, hydrogen, and oxygen atoms. They are classified based on the number of carbon atoms. The fats with 2, 3, and 4 carbon atoms are termed as short-chain, from 6 to 12 carbon atoms as medium-chain and more than 12 as long chain-fatty acids.

The SCFA are:

Acetate : 2 carbon atoms
Propionate : 3 carbon atoms
Butyrate : 4 carbon atoms

SCFA, like essential fatty acids, cannot be synthesized by human cells. Then why are they not included in the list of essential fatty acids? Because our friends, the tiny gut bacteria make them for us through digestion of dietary fibers. SCFA are also present in the diet, and as we know the saturated fats are present in animal fats, SCFA are present in butter and ghee, which is the same as clarified butter. Butter is a rich source of Butyrate from which it gets its name too. Why are these SCFA so important? We already get enough calories from the long-chain fats. Why do we need these short chain fats?

The SCFA are the predominant source of energy for the colocytes, which are the cells lining the walls of our large intestine. Thus SCFA provides energy to perform their function to the cells responsible for absorbing a large amount of water from our intestines. They keep them healthy and are essential for their proper functioning.

The SCFA butyrate performs a vital function in keeping us healthy as it keeps our weight in balance. How could it do that? Butyrate is a key metabolite that tells our brain that we are full while eating. Butyrate acts on intestinal cells, which then releases a compound Peptide YY. This Peptide YY reaches the brain and provides the satiety signal, basically tells that we are now full and should stop eating. In studies conducted on mice, long-term oral butyrate supplementation has been shown to reduce total food intake and prevent obesity and diabetes (Li et al. 2018). Butyrate also reduces the inflammation in our intestines by acting at the genetic level, thus reducing injury to the colon epithelial cells. This is how it prevents colon cancer, as longstanding injury to colon epithelial cells can ultimately lead to colon cancer (Hamer et al. 2008).

Propionate, another SCFA, is absorbed from the intestines and reaches the liver, where it promotes fatty acid oxidation and prevents the accumulation of fat in the liver.

Acetate, the smallest and most abundant of the SCFAs, can also act at the genetic level (Zeng and Chi 2015). It has been shown to reduce asthma and allergy in mice by working on immune cells. Even in humans, infants who lack the acetate producing gut bacteria are more likely to develop allergic diseases in childhood (Galazzo et al. 2020).

In short, the SCFA perform vital functions for our well-being. From being a source of energy to acting at the level of genes, they work in diverse ways. And the fact that the gut bacteria are their primary source makes us realize their vital importance. The gut-flora not only contributes to our genetic pool but can also act on our genes through the substances it produces.

MICROBIAL FEAR: BUSTING THE ANTIBACTERIAL MYTH

In the around mid-nineteenth century, we learned that disease –causing bacteria can be transmitted through water and food. The classic example of this was cholera. A British doctor John Snow investigated a cholera epidemic in London in 1854 and found that the source of out-break was a public water pump at Broad Street. The epidemic could be controlled simply by disabling that particular water pump. Thus, it was conclusively proven to the mankind that the food and water can contain bacteria that may cause diseases. The further credence to this finding was added by the work of other great scientists Louis Pasteur and Robert Koch during the second half of nineteenth century. Over the next century, this knowledge got augmented to a sense of fear with the discoveries of new bacterial or viral diseases. Food and water can indeed contain harmful bacteria. Still, the number of friendly bacteria far outweighs them (the number of disease-causing food or water-borne bacteria can be counted on fingers). And subsequently, due to that fear, we started to sterilize everything from UV treated water to pasteurized milk to even super-heated fruit juices. Not only that, we added chemicals in the form of preservatives about whose long term effects we did not know about. Some of these practices are essential for mass-produced and widely distributed products such as pasteurized milk but many other practices are not beneficial. By sterilizing the eatables due to our fear of microbes and the need to increase shelf life, we have devoid ourselves of the beneficial effects of friendly bacteria and exposed ourselves to the unknown long-term effects of added chemicals. We definitely need to follow the best hygienic practices, but we don't need to kill all the bacteria present in our environment.

Hygiene is a critical aspect of maintaining health. Handwashing can prevent many infectious diseases. But we need to understand that hygiene does not mean to get rid of all types of bacteria. "Do we even need antibacterial products to maintain hygiene?" It is a fundamental question. To answer this, let us look at a landmark study conducted in the field of public health in Pakistan. In the slums around Karachi, the rates of infectious diseases such as diarrhea and pneumonia were very high. A public health worker from Nebraska, US Stephen Luby, came to investigate the situation and offer possible solutions. He envisaged that if the people could wash their hands properly and regularly, these diseases can be prevented. He conducted a randomized controlled experiment taking a research grant from Procter and Gamble, who had manufactured a new anti-bacterial soap, Safeguard. The field workers distributed the soap to each household along with the instructions regarding when and how to wash hands. Half the households in the test neighborhoods were given the antibacterial soap, and the other half the regular soap. The control households were not given any soap. At the end of 1 year, there was a remarkable reduction in the number of infectious illnesses. The number of diarrhea patients was reduced by 52 % and pneumonia by 48 % in the 25 neighborhoods where the soap was distributed compared to the 11 neighborhoods where no soap was distributed. While this study confirmed the importance of handwashing, hygiene and taught about the principles of habit-change in the public health setting, the most startling observation was that there was no difference in infection rates from the households where antibacterial soap was distributed compared to those where normal plain soap was distributed (Luby et al. 2005). This again brings us to realize that while maintaining good hygiene is important, its beneficial effects are not increased by going antibacterial.

SECTION 2
GUT-FLORA AND LIFESTYLE DISEASES

All disease begins in the gut

Hippocrates

We, human beings or *Homo sapiens,* first originated from our ancestors around 200,000 years ago in Africa. We were just like any other species originating randomly according to the laws of natural selection. Well, just another animal, you see! We may have come a long way from our humble beginnings, but for most of our history, we have been hunter-gatherers. Agriculture originated during the Neolithic era (New Stone Age) only around 12 000 years ago. For a significant part of the history of our species, we have been collecting food roaming in the forests, either hunting animals or gathering plant products. This nomadic life has had important consequences related to our physical health. It ensured that our diet was diverse. In the early or middle Stone Age, a human being wandered through the forests, plucked and ate some fruits in the morning, hunted small animal(s) by the noon and gathered tubers to feed upon in the evening. Growing grasses, fruits or tubers to take care of the nutritional needs wasn't on the cards at all! Neither was rearing the animals or poultry for meat and eggs! Whatever we could get our hands on; we ate, depending upon the seasons and location. Most of these stone-age tribes would stay at a place for a season or a few and then migrated to new lands with different plant and animal-based foods.

In contrast, although today procuring food is a cakewalk making life super easy, this ease comes at a cost, a high cost, really! We have made massive progress in agriculture, storage, and supply chains enabling us to procure any food throughout the year. Whether we live in New York or a suburban European town or a village in India, we routinely eat similar kinds of food all through the year with wheat, rice, or corn as the staple cereal. Right, with no diversity whatsoever.

That was about the food; now, let's talk about physical activity. As hunter-gatherers, we had to perform tremendous physical labor to collect our food. Even upon the advent of agriculture, most of the human population worked strenuously in the fields to grow their food. But with the scientific and industrial revolution over the last 500 years, our need to perform physical labor to get foods on our platters has reduced dramatically and is at an all-time low since the past few years!

Although 500 years may seem like a huge time to us, it is a very short period on the evolutionary scale. Thus, although we have made great scientific progress in the last 500 years, our bodies have not changed as significantly. The body of a modern human adult responds to hunger in a manner similar to that of a 15th-century soldier, a 2000 BC peasant, or a tribal from the early or middle Stone Age. Throughout our history, we Humans have faced severe famines. Today, although we may have our refrigerators overflowing with scrumptious food, the body perceives starvation as the lack of food availability food, a stressful situation, similar to the times of famines that warrant minimizing calorie expenditure. This is the reason why the body lowers the metabolic rate when we starve.

Another factor, which has markedly evolved over the last few centuries, is stress. The life of our hunter-gatherer ancestors was not completely stress-free. But similar to the changes in food and physical activity, the stress has also changed, particularly the stressors. Let us again imagine a day in the life of our earliest ancestor. S/he woke up in the morning and began the search for the food. While roaming in the forest, he encountered a wild animal. His brain rightly perceived it as a danger and directed his adrenal glands to secrete epinephrine, which causes sympathetic activation and raised his blood circulation, increased the alertness and activity of his muscles. Armed with these beneficial responses, he ran to safety and saved himself. He resumed the search for his food and probably forgot about his adventurous encounter. He might or might not encounter another stressor in his day. Now compare it with our modern lives. The danger encountered by our ancestors was real but short-lived. While what we experience in our modern lives is continuous or *chronic stress*. Alas, our brain and our body respond in the same way to the stressors as that of our ancestors. As already stated, at an evolutionary level, we have not kept pace with our tremendous scientific and industrial progress. So when your omnipresent cell phone beeps with a notification, your brain responds to it similarly as that of a hunter-gatherer's to a predator. But unfortunately, you cannot run away from your cell phone. The continuous or chronic stress that a modern human encounters results in continuous activation of our sympathetic system and increased stress hormone levels, cortisol, which results in harmful effects in our body.

The lack of food-diversity coupled with physical inactivity and chronic stress has led to the emergence of a new class of diseases, popularly called the '*lifestyle diseases*'. Lifestyle diseases are the biggest killer of humankind today. Out of the top ten causes of death worldwide, four are lifestyle diseases, which include heart attack, stroke, diabetes, and cancers (WHO Global Health Estimates, 2016). Do gut bacteria have a role in lifestyle diseases? Can they protect us from them? Let us see.

Chapter 3

GUT-FLORA: DIABETES, OBESITY AND METABOLIC SYNDROME

Let us first talk about the importance of gut-flora in one of the most common lifestyle diseases, diabetes. Diabetes is characterized by an increased level of glucose in the blood. Glucose is an essential nutrient that is the chief energy source for the brain cells and the rest of our body. However, raised glucose levels in blood over a long period of time can have untoward consequences. It can damage the small blood vessels in kidneys and retina, leading to renal failure and vision-loss respectively. Besides, it also accelerates the process of plaque formation in the arteries, which is known as atherosclerosis. Atherosclerosis is the precursor to the most deadly lifestyle diseases; heart attack, and stroke.

Why do the blood glucose levels rise in diabetics? The blood glucose is regulated through the action of the hormones such as insulin and incretins which decrease glucose level in blood. On the other hand, hormones such as glucagon, corticosteroids, etc., do the opposite; they increase blood glucose levels. Consequently, diabetes can be caused by different mechanisms involving any of these hormones. However, the most common type of diabetes; Type 2 diabetes, occurs as a result of the reduced ability of insulin to carry out its normal function of facilitating the entry of glucose from the blood into the skeletal muscle and adipose tissue.

This inability is termed as 'insulin resistance'. One of the significant causes of insulin resistance is a high body-fat percentage. What is interesting to note is that the gut-flora in individuals with Type-2 diabetes differs from that of healthy individuals. According to a recent study, those with type-2 diabetes have lesser butyrate-producing bacteria and more Lactobacilli as compared to healthy individuals (Qin et al. 2012). Remember? Butyrate is one of the most important nutrients produced by gut bacteria. Our own human body - the epitome of creation, does not synthesize butyrate. Butyrate acts on the endocrine cells in the intestines to induce the secretion of peptide YY. Peptide YY travels to the brain from the intestines through the blood-stream. It tells the brain that we are full and should stop eating further. This 'feeling of fullness' is vital to prevent over-consumption of calories and the resulting obesity and diabetes.

Among other changes observed in the gut-flora are that Type-2 diabetes patients have significantly fewer Firmicutes, than the healthy individuals (Larsen et al. 2010). This study also found that the higher the Bacteroidetes to Firmicutes ratio the higher was the blood glucose level. Furthermore, Proteobacteria were also found to be more abundant in type-2 diabetes patients compared to healthy individuals. These observations suggest that Bacteroidetes and Proteobacteria may have a role in the pathogenesis of type-2 diabetes. Not only is the gut-flora different in Type 2 diabetics, it is surprising to know that it could also predict the future occurrence of diabetes. In a large population-based study conducted in Sweden, including more than 1500 subjects, researchers found that gut-flora was altered in patients with pre-diabetes. The reduction in butyrate production again being the most important finding (Wu et al. 2020).

We now understand that gut-flora is different between people with diabetes and healthy individuals. But does it mean that alteration in gut bacteria can lead to diabetes? Do we have scientific evidence? In a series of exciting experiments, fecal bacteria from lean and healthy people were transferred into the insulin-resistant individuals with metabolic syndrome. Six weeks after this fecal transplant from lean subjects, insulin sensitivity was improved in insulin-resistant individuals. On the fecal microbiota analysis, it was found that the number of butyrate-producing bacteria was increased in the gut-flora of the recipients (Vireze et al. 2012). This ability of gut bacteria to alter insulin resistance indicates that the gut bacteria have a definite role in the development of Type-2 diabetes, which is related to their ability to produce butyrate. Apart from butyrate production, there is another mechanism by which gut bacteria regulate blood glucose levels. To understand that, let us see what bile acids are. Bile acids are substances produced by the liver that are an important constituent of the bile juice. The bile juice is released into the intestines and helps digest the fats. The bile acids secreted by the liver are converted to secondary bile acids by the gut bacteria. These secondary bile acids further regulate the production of glucose in the body. The secondary bile acids act on a receptor called FXR and lead to its activation. This decreases the production of glucose in the body, thus preventing diabetes. The secondary bile acids also act on a class of intestinal endocrine cells, leading to the secretion of GLP-1. GLP-1 is an incretin hormone that leads to a decrease in glucose release from the liver into the blood, preventing diabetes.

All the scientific studies discussed above point towards one intriguing question- Can change in gut-flora composition cure diabetes? Well, so far, this has not been achieved. However, the studies have shown that changing the gut-flora with probiotics DOES increases the insulin sensitivity and also improves fasting blood glucose levels (Tao et al. 2020); but not to an extent to get the individuals off the medications, or to completely reverse the disease.

OBESITY AND METABOLIC SYNDROME

Obesity is both an independent disease, as well as a precursor to diabetes, cardiovascular diseases, fatty liver and bone, and joint-related problems. Obesity adversely affects almost every part of the human body and is a critical element of metabolic syndrome. Metabolic syndrome is a constellation of diseases such as obesity, hypertension, altered blood lipid levels, and elevated blood glucose. It indicates the presence of insulin resistance. We already know that insulin resistance is a phenomenon in which the insulin hormone secreted by the pancreas is unable to function effectively, resulting in increased blood glucose levels. Why is insulin not able to act despite its adequate secretion from the pancreas? In fact, in patients with Type 2 diabetes, insulin secretion is increased as compared to healthy individuals.

To understand this, let us first know about inflammation. It is now widely recognized that a state of persistent low-grade inflammation is the primary reason for the development of insulin resistance. The pro-inflammatory molecules which mediate the inflammation are produced inside the body by various immune cells, primarily macrophages. The most common pro-inflammatory substances are TNF-α, IL-1, and IFN-gamma. You must be wondering why our own body produces such molecules that can harm it. These pro-inflammatory substances are an indispensable part of the immune system, which protects us from infections by eliminating the pathogenic microbes from the body. Thus, a brief stage of inflammation is required for the effective functioning of the immune system. But a state of persistent inflammation is harmful to us as it leads to insulin resistance. These pro-inflammatory molecules cause insulin resistance by inhibiting insulin's action on skeletal muscle, adipose tissue and liver. These three organs are the primary sites for the uptake of glucose from the blood. And the inability of insulin to act effectively on these organs results in an elevated blood glucose level. Okay, but how are these pro-inflammatory molecules produced continuously, leading to persistent inflammation even in the absence of infection?

Well! The crux of this inflammation, insulin resistance, and metabolic syndrome phenomenon is a high-fat diet. A high-fat diet leads to increased levels of inflammatory mediators and circulating free fatty acids in the human body. There are multiple mechanisms by which a high-fat diet leads to a state of persistent low-grade inflammation, but it probably starts with the alteration of our gut –flora. Scientific evidence supports this hypothesis. In one such study, scientists fed a high-fat diet to the germ-free mice; our old friends, which, as you already know, are raised in a sterile bubble and do not have a gut-flora and the conventionally raised mice. Germ-free mice neither got obese, nor did they show evidence of increased inflammation, whereas the conventional mice showed both (Rabot et al. 2010). When germ- free mice were transplanted with gut microbiota from the obese mice, they showed an increase in body fat. This highlights that alteration of the gut-flora is an important reason by which a high-fat diet leads to obesity and persistent inflammation. Understood? But how does alteration in gut-flora leads to this persistent inflammation? Well, gut bacteria can regulate the level of inflammation in multiple ways. First, as we learned earlier, friendly gut bacteria play a huge role in the maintenance of an effective gut-barrier. An alteration in the gut-flora composition results in a compromised gut-barrier function, which increases the gut-permeability. This causes harmful pathogens/molecules/substances to enter inside the blood-stream and lymphatic channels from the intestinal lumen. These trigger the production of pro-inflammatory molecules such as TNF-α, IL-1, and IFN-gamma, causing persistent inflammation.

Secondly, obese individuals have a higher number of bacteria from the phylum proteobacteria in their gut-flora. The cell-walls or outer covering of the proteobacteria have lipopolysaccharide (LPS) as one of its constituents. The LPS is continuously released inside the gut as these bacteria die. This gut microbe-derived LPS is the key molecule causing persistent inflammation and insulin resistance. LPS itself is a potent inflammatory substance. Besides, the altered gut-permeability increases the amount of LPS gaining entry from the intestinal lumen into the bloodstream leading to a persistent low-grade pro-inflammatory state. Thus, alteration in gut-flora has a double impact on the balance of inflammatory status in our body (Figure 2).

The increased level of LPS in plasma is termed *endotoxemia*. Studies conducted in mice have shown that a 4-week high-fat diet can lead to increased lipopolysaccharide (LPS) concentration in plasma, which mediates inflammation. It increased the proportion of LPS containing bacteria in the gut-flora (Cani et al. 2007). In the same study, when endotoxemia was artificially induced in mice through a continuous subcutaneous infusion of LPS for 4- weeks, fasting blood glucose, insulin levels, weight, and body fat were increased to a similar extent as in high-fat diet-fed mice. All of these are simply the markers of insulin resistance.

It shows that endotoxemia can lead to insulin resistance. Thus, a *high-fat diet* leads to persistent low-grade inflammation through an alteration in gut-flora, which is a precursor to lifestyle diseases.

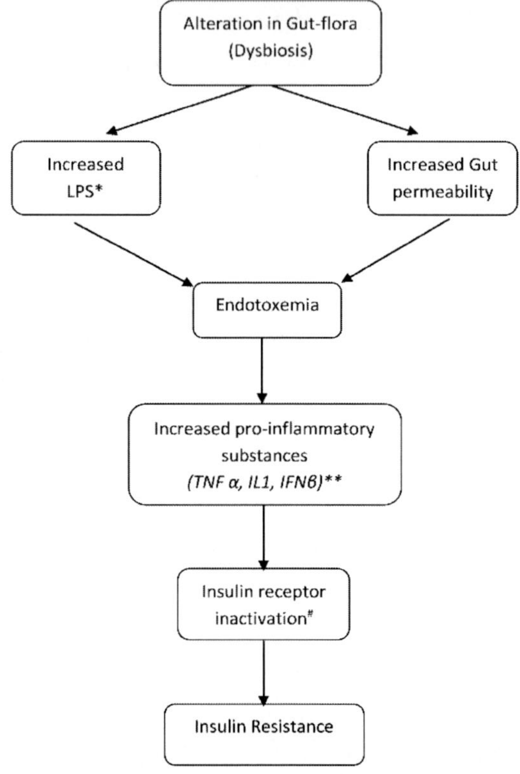

* Lipopolysaccharide.

** TNFα, Tumor necrosis factor α, IL1, Interleukin 1; IFNβ, Interferonβ.

Inflammatory markers promote phosphorylation of Insulin receptor Substrate (IRS1) at serine 307.

Figure 2. Alteration of the gut-flora leads to the development of Insulin Resistance.

Does the Obese Gut Differ from the Lean Gut?

We just learned about the biologically possible ways in which the alteration in gut-flora may lead to the development of obesity and metabolic syndrome. Let us now look at the composition of gut-flora in obese individuals. First and foremost, the gut bacteria in obese individuals lack diversity, which is a consistent observation across various studies on animal models as well as in humans. Obesity and insulin resistance are both associated with less bacterial diversity. Apart from low diversity, there is also an increased number of Firmicutes and decreased number of Bacteroidetes in their gut-flora leading to a rise in the Firmicutes/ Bacteroidetes ratio. An increase in Firmicutes probably leads to an increase in energy absorption from the food consumed (Balamurugan et al. 2010)). You may recall that in people with diabetes, it was reverse. Studies found people with diabetes to have reduced Firmicutes (Larsen et al. 2010). What is this 'Firmicutes paradox'? Does reduction in the population of Firmicutes in gut-flora lead to diabetes and their increase to obesity? To be honest, we are not sure yet. But Firmicutes probably held the key to this puzzle of diabetes, obesity, and metabolic syndrome. Besides, not all the species from the Firmicutes phylum (group) increase in individuals with obesity. Infact, one of the Firmicute bacteria, *Faecalibacterium prausnitzii,* decreases in number in obese individuals compared to the lean ones. Thus, while at the level of phyla (groups), Firmicutes increase in obese subjects, the situation may be reversed at the level of individual bacterial species.

Another bacterium that is reduced in obese and diabetic individuals is Bifidobacterium. Bifidobacterium is a vital member of our gut-flora. This bacterium is present in breast milk. Studies have shown that breast-fed infants have a higher concentration of Bifidobacterium in their intestines than non-breast-fed infants. According to the latest research, breast-fed infants not only resist the infections better but are also less likely to develop lifestyle diseases such as obesity and diabetes. And the secret ingredient in the mother's milk is none other than the lowly bacterium Bifidobacterium. Scientists have identified gut bacteria in children as young as six months old that can predict the tendency for future weight gain. They analyzed weight and other health parameters in these children until seven years of age. Again, the key regulator of weight gain turned out to be Bifidobacterium; the children with a higher number of

Bifidobacterium had lesser chances of being overweight at the age of 7 years (Kalliomaki et al. 2008).

Now we know that the gut-flora differs between obese and lean individuals. And it is an important player in the development of inflammation and insulin resistance. But if bacteria can lead to obesity, is it possible to transmit obesity by transferring the bacteria from an obese person to a lean one? Well! There are some interesting scientific studies that put this into practice and might just amaze you! A group of scientists transplanted microbiota from both the obese as well as the lean individuals to the germ-free mice. Any guesses what they found? Well, the total body fat was higher in the mice transplanted with obese gut-flora, even when the diet given to both the groups of mice was the same (Turnbaugh et al. 2006). For the first time in the history of medical science, obesity was seen in a new light as being a transmissible disease. Possibly the future sees obesity and diabetes as infectious diseases, the only caveat being that they are caused by a group of micro-organisms rather than a single microbe.

Fecal microbial transplantation (FMT) is the most ingenious way to change one's gut-flora. We shall read about FMT in detail in section3. Although the role and benefits of FMT are being investigated for a host of disorders, it is established as a therapy for the difficult to treat *Clostridium difficile* infection. *C. difficile* infects our large intestine and causes diarrhea, which can get pretty serious and even life-threatening. In an interesting chance observation, a patient with *C. difficile* infection who received FMT as treatment had an unintentional weight gain of 34 pounds in the months following the procedure despite a medically supervised diet and exercise program. Any guesses why it happened? Yes, this patient received the FMT from an overweight person. However, the gut-flora of this patient was not looked into for the changes associated with obesity (Alang and Kelly 2015).

If the gut bacteria can transmit obesity, can they cure it too? What happens if the gut-flora from a lean person is transferred to obese individuals? In a series of experiments, Dutch researchers transferred the gut-flora from a lean individual to obese men with metabolic syndrome by FMT. This led to a sharp increase in the insulin sensitivity of the recipients six weeks after the procedure. The gut-flora diversity was also increased suggesting that the microbial transplant altered the recipients' metabolism through the changes in their gut microbiome. However, the changes were not permanent as all the benefits were gone 18 weeks after the procedure (Koott et al. 2017). Another study conducted in Canada investigated the impact of fecal transplants on the liver fat content in individuals with fatty liver disease. The fecal transplantation could not reduce the fat content in the liver, but the intestinal permeability was reduced in those subjects in whom it was elevated before the procedure (Craven et al. 2020). In another human study, researchers in Massachusetts recruited 24 obese men and women with insulin resistance, along with four lean donors. They selected donors who had a history of always being very lean: people who said they could eat whatever they wanted and still remain skinny. Half the obese subjects took specially prepared frozen capsules containing stool from the donors on a weekly basis, while the others received a placebo. After 12 weeks, they found that the fecal treatment using capsules was safe and tolerable and that the subjects had acquired the gut bacteria that resembled that of their donors (Yu et al. 2020). But overall, unlike in the Dutch study, there was no improvement in their metabolic health.

So finally, what does all this scientific evidence say? To summarize, alteration in the gut-flora can lead to the development of diabetes and metabolic syndrome. Perhaps the gut-flora plays a central role in a large number of patients suffering from these ailments. But can we use the gut bacteria to cure these diseases? Not yet. We shall definitely be there in the future, but right now, we have to wait.

Chapter 4

Gut-Flora: Stress and Behavior

We all experience stress. It may be trivial (the omnipresent traffic jams, the important task that was due last week, the grocery shopping you must do today), more serious (arranging finances for your own home, getting acclimatized to a new work-place) or persistent (caring for an ill parent, being unemployed); each of us faces stress from time to time. Depending on the stressor's severity, how we respond to it, and how long it lasts, it can affect our bodies and brains differently. A short-term stressor, like getting caught in the traffic, might increase your blood pressure and have you feeling annoyed for a few minutes whereas a more chronic stressor that lasts weeks or months may precipitate depression and anxiety in some individuals.

We learned how our brain and gut could communicate with each other in chapter 2. You may recall that there are neural, immune, and endocrine connections through which this communication occurs. And together, they form the Gut-Brain axis. But is there a relationship between stress and gut-flora? Does stress influence gut-flora? Or can gut-flora impact our mood? These questions must be racing in your mind. Let us find the answers.

Does Stress Impact the Gut-Flora?

Yes, stress has a profound impact on our gut-flora. Stress may even have a lasting effect on the gut-flora if it occurs early in life. Early life stress is particularly a serious form of stress. It is also called as adverse childhood experiences or "ACEs". This includes abuse, neglect, or a chaotic home environment during childhood or teenage years. Early life stress is particularly important because it can increase the risk of many illnesses in the future, including depression and heart diseases, even if the person has overcome the stressful environment of their childhood or adolescence. There are no definite reasons for it, but it probably occurs because the stress has occurred during a sensitive period of brain development when our stress-response is developing and disruptions at this time point can set one up for vulnerability later in life.

To understand the effect of stress on gut-flora, we would first examine the studies conducted in mice. In one such study, mice were separated from their mothers daily for 3 hours soon after birth to mimic an early life stress situation. The mice that were separated from their mothers had an altered gut-flora as compared to the non-separated mice, particularly with reduced bacterial diversity. Besides, these isolated mice had higher levels of stress hormones and increased pro-inflammatory substances (O'Mahony et al. 2009). Yes, the same pro-inflammatory substances which cause a state of persistent low-grade inflammation- a precursor to lifestyle diseases. Taking a cue from this study, researchers have conducted studies in humans where they have compared the gut-flora of the individuals who experienced early life stress and compared that to individuals without such stress. In a study conducted in healthy pregnant women, half of whom had experienced multiple childhood stressors, researchers found that the women who had stress in early life had a different gut microbiota than women without such stressful experiences. These women were both physically and psychologically healthy at the time of study, but still, the early life stress they experienced had an impact on their gut-flora even years later. Specifically, women with high early life stress had elevated gut *Prevotella* compared with women without early life stress. This is interesting, as studies have shown that this bacterium is associated with inflammation and altered cortisol response to acute stress. And we know that elevation of both these factors harm our health. Not just that, you would be surprised to know that maternal stress during pregnancy can even affect the gut-flora of the baby. Infants born to the mother who reported high stress during pregnancy have shown to have lower numbers of Bifidobacteria- our most important friend during infancy and early childhood (Zijlmans et al. 2015).

In another human study, looking at the role of early life stress on gut-flora, scientists examined the microbiota of teens who had lived in an orphanage in early life (a form of severe chronic stress) and compared it with the microbiota of children who had been raised in a stable family environment. When these children were studied at around age 10–15, they had altered gut microbiota compared with the children who had spent their entire lives in a stable family environment (Callaghan et al. 2020). Together, these findings tell us that early life stress can impact the gut-flora during childhood, which can persist through adolescence and adulthood. At least some of these changes, such as the increased presence of Prevotella, can predispose us to lifestyle diseases. You would be surprised to know that stress affects the gut bacteria in the same way as high-fat diet. In a study conducted on mice, scientists found that gut-flora changes in response to induced stress and high-fat diet were similar (Bridgewater et al. 2017).

How Gut-Flora Affect Stress-Response?

You may recall that the effect stress has on us is determined by the severity of stress, its duration, and last but not least, how we respond to it. In-fact, psychologists, lifestyle coaches and mindfulness experts believe that stress-response is not just the most important factor determining how well we come out of a troublesome situation but can even be manipulated. Obviously, gut bacteria cannot remove the stress from our lives or alter its severity and duration. (If they could do so, they would be the GOD). BUT, can they change the way we respond to stress? Let us see!

What is stress in biological terms? In the most simplistic explanation, stress is an event that leads to the secretion of certain hormones inside our body, mainly cortisol and epinephrine. These hormones produce effects such as an increase in heart rate and respiratory rate, and alertness, which we may perceive as palpitations, breathlessness, and difficulty in sleeping, respectively. We must understand that all these responses are beneficial to us in the short-term and thus they are ingrained in our system and cannot rid ourselves of them. When as hunter-gatherers, we encountered a wild animal, these responses gave us increased alertness and provided more oxygen to our muscles so that we could save ourselves. But the continued stress, we modern humans experience due to our lifestyles lead to the persistent elevation of cortisol in our bodies, which is extremely harmful.

The adrenal gland in our body secretes cortisol. The process of cortisol secretion starts with the registration of stressful events by our brain. The brain sends a signal to the adrenal gland in the form of ACTH hormone, which directs the adrenal gland to secrete cortisol. This axis, which is involved in the secretion of cortisol, is known as the hypothalamic pituitary adrenal (HPA) axis. Gut bacteria are important because they are essential during the early developmental stage for a fully functional HPA axis. It has been proved in an elegant scientific study. To study the impact of gut-flora on the development of the HPA axis and its response to stress, researchers measured ACTH levels and cortisol in germ-free mice and conventionally raised mice. They found that when stressed, both these hormones were higher in germ-free mice. They enriched these germ-free mice intestines with a friendly bacterium and found that this heightened stress response was reversed. Now, the same germ-free mice had similar cortisol levels compared to the wild mice in response to stress. And this friendly bacterium was none other than Bifidobacterium, which is abundantly present in the gut of breast-fed infants. Now, the most important finding in this study was that such reversal of heightened stress response of germ-free mice could be corrected only at an early stage, which therefore indicates that exposure to bacteria at an early developmental stage is required for the HPA system to become fully susceptible to inhibitory neural signals (Sudo et al. 2004).

In another meticulously conducted study, the researchers at Cornell University in New York, US, started with two groups of mice – one healthy and the other treated with antibiotics to wipe out their gut bacteria. They put the mice in a special chamber where a tone sounded for 30 seconds. When the tone stopped, the mice got an electric shock to their feet. As you might expect, the mice started freezing with fear as soon as they heard the tone in anticipation of electric shock. However, mice can be trained to unlearn such kind of fear. To make them unlearn this fear, this time the tone was sounded but without any shock. After a few sessions of tone without the shock, mice were expected to stop freezing when they hear it. This is what happened to the normal mice. But the antibiotic-treated mice just could not stop freezing to the tone, long after the shocks had stopped (Chu et al. 2019). These antibiotic-treated mice with deficient gut-flora just could not learn to extinct their fear, which happens to us so commonly. How many of us keep getting troubled by a stressful event in the distant past, finding ourselves unable to get over it? The answers may lie in our gut-flora. But how could the gut bacteria impact our stress-response? To answer this, the scientists in the same study first cut the Vagus nerve, which supplies the information from our gut to the brain, but it made no difference to the unlearning of fear. They did not find any difference related to the immune system too. They ultimately found that the substances produced by gut bacteria present in the cerebrospinal fluid that surrounds the brain were markedly reduced in the mice who couldn't unlearn the fear. Thus, the relation between the gut bacteria and our ability to forget the fearful experience seems to be direct, related to the metabolites produced by the bacteria themselves rather than anything else.

Role of SCFA (Yes, again) and Other Bacterial Substances

Now we know that gut bacteria can influence our response to stress due to the metabolites produced by bacteria rather than anything else.

Could these bacterial substances alter the way we respond to stress? To answer this, let us look at an experimental study conducted at University College Cork, Cork, Ireland. In this study, the researchers took two groups of mice. The first group was fed SCFAs for one week, and the other was not. After that both the groups underwent three weeks of psychosocial stress, which was followed by evaluation. The stress was delivered by keeping the individual mouse with a larger aggressive mouse every day for a specified period. The SCFA fed group had lower levels of stress hormones (cortisol) as well as showed reduced stress-related behavioral changes as compared to the non-SCFA fed group (Van de Wouw et al. 2018). We know that the SCFA are exclusively produced by gut bacteria inside our body. This reaffirms that gut-flora have a huge role in deciding the way we respond to stress. In fact, gut-flora might just be the KEY to happiness. Serotonin, also known as HAPPY HORMONE, is a neurotransmitter that serves many functions in the human body, including playing a role in our mood. Serotonin levels are reduced in depression and anti-depressant drugs SSRIs act by increasing serotonin's availability to the brain cells particularly in the regions that deal with our emotions. Now you must know that more than 90 percent of our body's serotonin is synthesized in the gut from an amino-acid tryptophan. And, of course, the gut bacteria play a significant role in it. In a study conducted in mice, the researchers at UCLA found that a host of substances produced by the gut bacteria, particularly Clostridia, interact with the serotonin-producing cells in our large intestines and promote its production. Fifty percent of gut serotonin production is due to the effects of these metabolites produced by the gut bacteria. When scientists infused these bacterial substances in the colon of germ-free mice, it led to increased colonic and blood serotonin levels. These substances include deoxycholate (a secondary bile acid), α-tocopherol (Vitamin E), and tyramine (Yano et al. 2015). In *Sapiens- A Brief History of Humankind,* Yual Noah Harari writes about happiness and discusses happiness from the biochemical, philosophical, social, and spiritual points of view. However, in the future, any discussion about happiness would be incomplete without including the *gut-flora* dimension.

GUT-FLORA AND DEPRESSION

One in five individuals suffers from depression at least once in their lifetimes. It is a mental illness characterized by the sadness of mood, low energy levels, inability to derive pleasure from previously pleasurable activities, sleep disturbances, memory deficits, hopelessness, worthlessness, and helplessness. In extreme cases, it may also lead to suicidal ideation. The studies show that gut-flora composition is altered in individuals with depression. In a large Belgian study, including more than a thousand volunteers, the gut-flora of depressed individuals was found to lack two specific bacteria; both of them produce butyrate (Valles-Colomer et al. 2019). In an exciting discovery, the gut bacteria have been shown to be directly linked to depression in humans. The bacteria Oscillobacter is found to increase in number in people with depression. Valeric acid, a metabolite produced by Oscillobacter, acts as a natural tranquilizer. Valeric acid mimics the action of GABA, a neurotransmitter that calms the activity in the brain and can lead to depression. Increased Isovalreic acid levels in stools have been shown to correlate with depression in humans (Szczesniak et al. 2016). Other bacteria that are commonly found to be associated with depression are Alistipes. Alistipes interfere with the availability of amino acid tryptophan in the intestines. You know that tryptophan is the raw material for the production of the happy hormone Serotonin; thus, the increased abundance of Alistipes can reduce serotonin levels and lead to depression (Jianh et al. 2015).

Now the question arises if the alteration in gut-flora can cure depression? A study conducted in 2008 reported increases in tryptophan levels in rats treated with the probiotic *Bifidobacterium infantis* (Desbonnet et al. 2008). Many studies with various friendly bacteria such as Bifidobacterium have shown benefit in depression resulting in improved mood. (Wallace and Milev, 2017). However, at present, the effects shown are modest and, thus, probiotics are not routinely prescribed for the treatment of depression.

Gut-Flora and Behavior

Behavior is the way we conduct ourselves, the way we interact with others. It is what we are. Gut bacteria can determine the way we respond to stress but isn't it scary if they can dictate even our personality? We already know that we are only 10 % human at cellular level and only 1 % of genes we carry are our own. But it is difficult to believe that the way we respond to everyday events, and how we go on in our lives can be influenced by our gut-flora. Is it true?

In a study published in 2011, scientists gave BALB/c mice, a particular strain of mice that are typically timid and shy, a cocktail of antibiotics, which changed the composition of their gut bacteria. And it also changed their behavior. From being timid and shy, they become bold and adventurous (Bercik et al. 2011). When the antibiotics were stopped, the mice reverted to their usual selves. The researchers followed this up with another experiment. This time along with the timid BALB/c mice, they took NIH Swiss mice, known for their courageous, exploratory behavior. The researchers then colonized each group of these mice with bacteria from mice of the other strain with opposite behavior. Could you guess what happened? The normally timid BALB/c mice became much more fearless explorers, while the typically adventurous NIH Swiss mice suddenly grew more hesitant and shy. These results underscore, that at least in laboratory mice, some behavioral characteristics are driven not solely by the animals themselves but also by bacteria inhabiting the gut. Whether this holds true for humans whose gut-flora as well as behaviors are much more complex remains to be seen. Though a recent study by Kim et al. conducted in 672 adults showed a correlation between gut bacteria and personality traits in humans, such a direct behavior change with the alteration in gut bacteria as seen in a study by Bercik et al. may never get demonstrated (Kim et al. 2018). Kim et al. found that individuals with high conscientiousness showed an increased abundance of butyrate-producing bacteria, including Lachnospiraceae. These findings offer definite insights in the relationship between gut-flora and behavior.

Gut-Flora and Social Interaction

One of the behavioral aspects in which the role of gut-flora is firmly established in humans is social interaction. We, humans, are social animals. We thrive on our interaction with others. Among all the life forms existing on our planet, we are the most socially active creatures. How do we know that gut bacteria influence the social aspect of our personality? First of all, germ-free (GF) mice have shown to be less socially active than conventionally raised mice. Desbonnet et al. examined social behavior in GF mice using the three-chamber sociability test. In the study, a mouse was placed in the middle chamber with a second mouse in the first chamber and an object in the third. A conventionally raised mouse spent more time with a second mouse than with the object, while the GF mouse spent far less time with another mouse and a greater amount of time with the object. In the second experiment, the mouse was placed in the middle chamber with a familiar mouse in the first chamber and an unfamiliar mouse in the third chamber. The conventionally raised mouse spent more time with the unfamiliar mouse, while the GF mouse showed no preference for either the familiar or unfamiliar mouse. This study clearly demonstrates that GF mice have a reduced ability to interact with their surroundings (Desbonnet et al. 2014). Furthermore, the introduction of the gut bacteria in GF mice reversed some of these symptoms.

Autism Spectrum Disorder (ASD) is a group of neurobehavioral syndromes characterized by deviations in social interaction, poor nonverbal communication, and speech deficits. It may manifest as a mild syndrome correctable by training or a severe untreatable syndrome. It presents as poor eye contact, repetitive stereotypical movements, and speech difficulties. Autism usually presents in early childhood, with symptoms starting between the age of 12 to 18 months in most children. Studies on gut-flora in children with ASD provide us rare insight into the role gut bacteria play in determining our behavior. Autism has been found to be associated with diarrheal episodes, gastrointestinal symptoms, and use of antibiotics during the 1st year of life (Adams et al. 2011). Several studies have shown that children with autism have different gut microbiota than their siblings without the disease. . They have been shown to have reduced levels of different bacteria, including our old friend Bifidobacterium. Such findings led the researchers at the California Institute of Technology to transfer the gut bacteria from children with autism to germ-free mice. The offspring of these mice socialized less as compared to the offspring of the mice transferred the gut-flora from donors without autism (Sharon et al. 2019). Thus like obesity, autism could also be transferred to another organism with the transfer of gut bacteria. In the same study, researchers identified substances produced by gut bacteria that were reduced in the autistic mice (amino acids taurine and 5-aminovaleric acid), and their supplementation led to an improvement in the autistic behavior and increased social interaction.

Not only in the children, but the gut-flora has also been shown to impact social behavior in adults too. In a recent study conducted at Oxford University, researchers analyzed the gut microbiome of 655 individuals, along with questionnaire-based information about their personality, behavior, and sociability. They found that people who have larger social networks (no, they don't mean a large number of Facebook friends) are more likely to have greater gut bacteria diversity. Furthermore, specific bacteria such as *Desulfovibrio* and *Sutterella* were found to be reduced in people with increased sociability (Johnson 2020). Both these bacteria have shown to be increased in children with autism. Let us put the gut-flora through the litmus test again. Does change in gut bacteria lead to an improvement in autistic disorders in humans? Studies have utilized both antibiotics and probiotics to effect a change in gut-flora in children with autism. In a study with poorly absorbed antibiotic oral vancomycin, eight out of the ten children showed improvement in their symptoms, but the benefits waned off over a period of two to eight months after stopping the treatment (Sandler et al. 2000). As far as probiotics are concerned, a large number of studies have shown mixed results. However, a recent systematic review that included all the randomized trials concluded that there is limited evidence to support the role of probiotics in treatment of autism (Ng et al. 2019). As in other diseases related to gut bacteria, the microbial transplantation is also being studied in autism especially in patients with associated gastrointestinal problems. It has shown improvement in gastrointestinal as well as autism symptoms (Kang et al. 2017). Although we have gained tremendous knowledge about the role of gut bacteria in autism but are still in the process of learning how to utilize it for our benefit. As treatment for autism particularly remains challenging, the success of gut microbial therapy would be life-changing for so many children and young adults.

Apart from behavior and social interaction, gut-flora affects our memory too. It has been found that the germ-free mice have absent working memory and decreased brain-derived neurotropic factor (BDNF) in the hippocampus (Gareau et al. 2011). Hippocampus is the part of the brain which deals with memory while BDNF helps in the growth and development of our brain cells. Besides, the elderly who suffer from Alzheimer's have been shown to have less diverse gut-flora compared to those with good memory (Claesson et al. 2012). If you have been struggling to understand someone, it is probably time to shift your gaze from the brain to their gut.

Gut-Flora in Irritable Bowel Syndrome

If gut-flora can influence the working of our brain, have an impact on metabolic disorders, it is obvious they play a role in the diseases of the gut. We have already learned that gut bacteria provides protection against harmful bacteria by fighting with them for the food and space inside the intestines and thus prevent gut infections. Gut-flora plays an important role in a specific group of disorders characterized by inflammation in the intestines and is termed as inflammatory bowel disease. It is now established that this occurs due to an abnormal response to the gut-flora by our immune system, which leads to increased inflammation and harms the gut itself. It may be due to genetic factors or alteration in gut-flora composition itself, which triggers the immune system and leads to injury. However, the intestinal disorder that is more common and affects many people is irritable bowel syndrome (IBS). IBS is amongst the type of diseases commonly regarded as *functional gastrointestinal disorders*. They are characterized by the gut-related symptoms but without any apparent abnormality. People suffering from IBS have pain in the abdomen and an associated change in their bowel. It is commonly noticed that these symptoms increase during the period of stress and anxiety.

How do these symptoms occur without any identifiable cause? In-fact these disorders reinforce the link between our brains and the gut that they can communicate with each other. The brain has a direct effect on the stomach and intestines. For example, the very thought of eating can release the gut digestive juices before food gets there. Therefore, a person's intestinal distress can be the product of anxiety or stress. This is especially true in diseases such as IBS, where a person experiences gastrointestinal upset with no obvious cause. As we know the gut-flora is an indispensable part of this communication channel between the gut and the brain. It plays an important role in the genesis and the treatment of functional gastrointestinal disorders. Many scientific studies have demonstrated the alteration in gut-flora in patients with IBS (Dupont 2014). Manipulation of gut-flora by probiotics improves the gastrointestinal symptoms and reduces the anxiety in IBS patients (Williams et al. 2009).

We could see how the most common lifestyle diseases, whether diabetes, obesity, and metabolic syndrome, or stress-related disorders, are affected by the alteration in the gut bacteria. Obesity and stress are the two most important and widespread diseases, affecting young as well as old. Both of them predispose to a large number of other diseases. And the small tiny bacteria present in our intestines play a definitive role in the genesis of obesity and the way we respond to stress. Gut-flora affects both our physical and mental well-being. They are our closest and the oldest friends being with us from the time the first human stepped on earth. Don't you think it is essential to know about them! Thank them! As they say, a combination of power and ignorance is a disaster. We are more and more powerful in our fight against bacteria but let us get rid of the ignorance about them before it leads to a colossal disaster. Let us embrace what is always inside us.

Chapter 5

HEALTHY GUT-FLORA - KEY CONCEPTS

What is healthy gut-flora? Well, by now, you must have a fair idea about it. Amidst all the scientific jargon and technicalities, the concept of healthy gut-flora is fairly simple.

While learning about the healthy gut-flora, let us first know about a few characteristics of gut-flora and some commonly used terminologies.

The gut microbiota has some inherent properties, which are:

1. Stability

   As we learned, that gut-flora starts developing soon after birth and changes rapidly during initial years of life depending on weaning, exposure to different types of foods and other environmental factors. At the start, it is simple and less diverse but becomes highly complex during early childhood. After that, during our adult life, the composition of the gut-flora remains relatively stable. Although our gut-flora changes with so many factors, as already explained in chapter 1, but if we study the gut bacterial composition of an individual at different time points, a large proportion of gut bacteria remains the same. This does not mean that we cannot change our gut-flora to achieve a healthier state. It means that changes in lifestyle can improve the gut-flora but we simply have to be persistent in our efforts.

2. Plasticity

The property of plasticity means that although the gut-flora is stable, on this background of stability, there is a continuous change in its composition and metabolic behavior depending on the factors such as diet, exercise, use of medications, etc. It is easy to understand as we have already seen how gut-flora changes in response to a large number of factors. What we must realize is that the property of plasticity gives the key to the healthy gut-flora in our own hands. We improve our diet, and our gut-flora will improve in response. We start exercising, and in turn, the gut-flora gets healthy. But it also means that we have to be constantly vigilant in our efforts as if we start eating unhealthy or become sedentary, the gut-flora will sense it and change itself accordingly.

3. Resilience

The gut-flora is resilient, which means that it has an inherent property to return to its steady-state. It tries to preserve its composition. Let us understand this with an example. A common insult to which every human's gut-flora is exposed is antibiotics. If you take an antibiotic for an infection, it will alter your gut-flora because it will kill some of the friendly bacteria present in your intestines. But, after you are cured and stop taking the antibiotic, the gut-flora slowly reverts to its original composition. It may take up to 1 month or even longer, but gut-flora always tries to go back to its original composition.

What are the implications for us if the gut-flora is resilient? It helps us in achieving the healthy state of gut-flora if we have to subject it to brief unavoidable harm such as antibiotics. It also means that to change our gut-flora to a healthier state, we must work continuously. We have to adopt the lifestyle measures permanently. If you missed, the gut-flora would go back to its initial composition. This also means that adopting lifestyle measures that can be sustained for life-long is a better way to achieve a healthy gut-flora than some quick -fix solutions such as enemas or medications such as probiotics.

If you look at these three qualities of the gut-flora, you might notice that these are inter-related (figure 3). The gut-flora remains stable as a whole, but changes in response to external factors to variable degrees and tries to come back to its original composition when that factor is removed. These properties also give us basic principles to achieve a healthy gut-flora; persistence and vigilance. We need to be persistent with our efforts and be always vigilant to avoid unnecessary harm to the gut-flora.

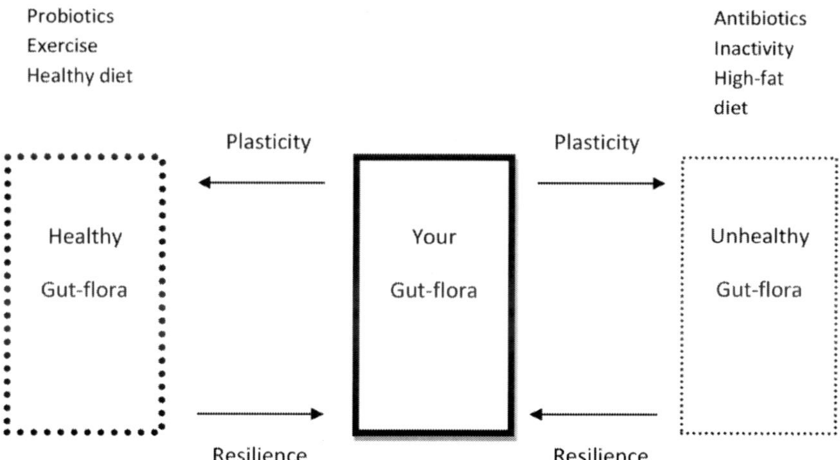

Figure 3. Concept of plasticity and Resilience of Gut-flora. Gut-flora shows plasticity i.e., its composition changes under the influence of various factors. But as that factor is removed, it goes back to its natural state showing resilience.

Now, what a healthy gut-flora means? We looked at the gut-flora from the point of various lifestyle diseases ranging from obesity and diabetes to stress. We also learnt about the factors such as diet, exercise, effect of medications, aging, etc. that influence gut-flora. Although some of the scientific studies we discussed showed an association of a particular bacterium to a disease while the others showed abundance or reduction of a phylum (group) related to a disease. But the one factor which is a common thread is *DIVERSITY*. Across the studies and diseases states, the most common observation is that the individuals who are not healthy have a REDUCED DIVERSITY of their gut-flora. Based on what we have learned so far, a healthy gut-flora has the following characteristics:

1. Diversity
2. Firmicutes: Bacteroidetes ratio
3. Less Proteobacteria
4. Abundance of Bifidobacteria

DIVERSITY

A healthy gut-flora is diverse, which simply means it has a large variety of bacteria.

Let us understand with an analogy. Imagine that there are two lakes in the forest. Lake 1 has ten different fish species, and Lake 2 has 100 different species of fish. It is clear that Lake 2 has a more diverse fish fauna. Similarly, a gut with a large number of different species of bacteria has diverse gut microbiota.

One of the most common parameters used to indicate the diversity of species at any place is *Simpson's Diversity Index.* Its value varies from 0, which means no diversity, to 1, which means infinite diversity. Although originally devised for ecological studies, Simpson's index is widely used to indicate the diversity of gut-flora. It is calculated by a formula that is beyond the scope of this book. Let us just remember that Simpson's Index is one of the ways by which we can indicate the diversity of gut-flora in an objective way.

The other terminologies which are commonly used in studies, books, and articles about the gut microbiome are *alpha-diversity* and *beta-diversity*. However, they appear complicated but are reasonably simple to understand. Let us go back to our Lake example. Lake 1 had ten different species of fish, and Lake 2 had 100 species. When we talk about each lake individually, we are talking about alpha-diversity. For example, for the purpose of understanding, we can say the alpha-diversity of Lake 1 is 10 and of Lake 2 is 100 which means the fauna of Lake 2 is more diverse. But when we compare the different fish species found in the two lakes, then it is termed beta-diversity. In the current example, if all the ten species found in Lake 1 are different from the 100 species found in Lake 2, then it means that the beta diversity between them is high.

In the context of the gut microbiome, when an individual's gut-flora is being described as diverse or less diverse, we are talking about alpha-diversity. And when the gut-flora of two individuals or different populations is compared with each other, it means we are talking about beta-diversity. If the Indian gut microbiome differs markedly from an American, it means there is high beta-diversity between the two. At the same time any one of them may be having high alpha-diversity depending upon the number of different bacteria species they contain. It is paramount to note here that it is the alpha-diversity that matters for an individual's point of view. What we have been talking about in this book is alpha-diversity. When a study says that the gut microbiome of diabetics is less diverse as compared to non-diabetics, it is talking about alpha-diversity.

Overworked!! Well, the above discussion is to just make your way smooth if you ever read a gut microbiome study. It will help you to understand them better. If this book is the only thing you are ever going to read about gut-flora, you can simply move ahead without bothering yourself. But please remember, *the healthy gut-flora is diverse.*

FIRMICUTES -BACTEROIDETES RATIO

We know that Firmicutes and Bacteroidetes are the two most common phyla in our gut-flora. An alteration in their number and composition is seen in a large number of diseases. Individuals with obesity are shown to have an increased abundance of Firmicutes or a higher Firmicutes-Bacteroidetes ratio. Higher Firmicutes-Bacteroidetes ratio has even been considered as the hallmark for obesity. Does an increase in Firmicutes means increased weight? Would eliminate them lead to weight loss? The gut bacteria are so complex, and our understanding about them is at best infantile that such generalizations usually do not hold true. If you recall, the studies in diabetic individuals found Firmicutes to be reduced in the gut-flora. Not just that, a higher Bacteroidetes-Firmicutes ratio was associated with higher blood glucose levels (Larsen et al. 2010). Even in obesity, many other studies have shown contradicting results, including a decrease in Firmicutes (Magne et al. 2020). Even when studies show an increased abundance of Firmicutes, a closer look has shown that bacterium such as *Faecalibacterium prausnitzii,* which itself is a Firmicute, is reduced (Balamurugan et al. 2010). Please remember, Firmicutes is the name of a phylum or group of bacteria. It includes a large number of different types of bacteria such as Lactobacillus, Faecalibacterium, etc. Apart from that, the abundance of Firmicutes is increased on consumption of high fiber diets associated with reduction in weight and insulin resistance. Besides, a reduction in the diversity of gut bacteria is a more consistent finding across the studies in obesity as well as diabetes irrespective of changes in the Firmicutes.

PROTEOBACTERIA

An unhealthy alteration in gut-flora is termed as *dysbiosis.* Proteobacteria are considered markers of gut dysbiosis. Proteobacteria, like Firmicutes, is a group of bacteria. They have a low abundance in the gut of healthy humans. Most of the pathogenic bacteria which cause intestinal infections belong to this group. Even though not all of them are pathogenic, their increased abundance has consistently been associated with increased inflammation. We know that mild chronic inflammation is the precursor to insulin resistance and metabolic syndrome, ultimately leading to diabetes and obesity. The Proteobacteria have cell-walls that contain lipopolysaccharide (LPS). LPS gets released in the intestinal lumen from these bacteria and gets absorbed and lead to increased inflammation. LPS binds to toll-like receptors (TLR-4) and leads to the activation of various pathways that increase pro-inflammatory substances inside the body (Boutagy et al. 2016). It is no coincidence that at both the extremes of age we have abundance of Proteobacteria while the strong, healthy and diverse gut-flora keeps them in check when we are young adults.

ABUNDANCE OF BIFIDOBACTERIA

Bifidobacterium is a member of phylum Actinobacteria, the fourth most abundant group in gut-flora. Bifidobacteria abundance in gut-flora has been associated with many beneficial effects. The primary reason for this significance is the stage of our life during which Bifidobacteria predominates. It is one of the first bacteria to colonize our intestines and has increased abundance in breast-fed babies (Bezirtzoglou et al. 2011). As infancy is a critical period in the development of the immune system, nervous system, and hypothalamic-pituitary axis, an alteration in gut-flora during this early period of life has long-lasting consequences. The abundance of Bifidobacterium during infancy is an indicator of healthy gut-flora. Its presence is associated with fewer infections and reduced chances of obesity during childhood (Kalliomäki et al. 2008).

OKINAWA: THE LAND OF IMMORTALS

Okinawa, an island in Japan, is the healthiest place in the world. There are 24.55 people over the age of 100 for every 100,000 inhabitants in Okinawa, which is far greater than the global average. A village Ogimi, situated on the northern end of the island, has the life expectancy in the world and is nicknamed the Village of Longevity. We are striving to know about the healthy gut-flora. Why not look at the gut of the healthiest people on the planet? In a study comparing the gut microbiome of healthy Japanese people with that of individuals from 11 other countries, striking differences were observed in the Japanese gut-flora. The phylum Actinobacteria was most abundant in the Japanese gut-flora, much more than individuals from any other country. Actinobacteria accounted for an average of 16% of bacteria in the gut of Japanese people. Now you know that Bifidobacteria belongs to the phylum Actinobacteria. Would you be surprised to learn that Bifidobacterium was the most abundant species in the Japanese gut-flora? Probably not! But! Yes, it is the most abundant bacterium in the Japanese guts. When the researchers analyzed the Japanese gut-flora functions, they found the high capacity to metabolize carbohydrates resulting in increased production of short-chain fatty acids (Nishijima et al. 2016). Thus, the guts of the world's healthiest people have abundant Bifidobacteria and an amazing capacity to produce SCFAs.

Do You Need to Know about Your Gut-flora?

Finally! We are here. After knowing about the gut-flora, its role in keeping us healthy and what is a healthy gut-flora, these thoughts must be arising in your mind. How does my gut-flora look like? Which bacteria are predominant in it? Whether it is the beloved and kind Bifidobacteria or the much-vilified Proteobacteria? Should I even attempt to know about it?

Well. Let us know a little about the process of mapping the gut microbiota. What you need to do is just collect your stool sample and hand it over to the collection guy. But it is a very advanced science that incorporates the latest in the DNA sequencing technologies and data management systems. The majority of bacteria present in our gut cannot be studied by traditional culture methods on an artificial media as it is not possible to provide the ideal and highly complex environment they get in our intestines. To identify these bacteria, scientists look for their genome in the sample. First, they extract DNA from the stool sample you provided. Then the DNA is subjected to sequencing using any of the current technologies such as Next Generation Sequencing. And finally, these sequences are matched with the available databases to identify the bacteria. This way, we get to know the composition of the gut-flora. However, researchers can dig deeper and look for the actual genes and the metabolites, which are the final products of these bacteria. For example, in the above-mentioned study on the Japanese gut microbiome, scientists looked for the metabolites and found the increased production of SCFAs in Japanese individuals.

If you are a healthy person, there is no need for you to know the composition of your gut microbiome. Likewise, suppose someone suffers from a disease in which there is definite and proven role of microbial therapy such as *Clostridium difficile* infection. In that case, s/he can straightaway undergo treatment without the gut microbiome analysis. Now we are left with a plethora of lifestyle diseases, stress, and behavioral disorders in which gut bacteria play a definite role. You must understand that although an alteration in gut-flora is a part of these illnesses, we don't yet have safe and effective microbial therapies for them. Even if they exist, they are not approved by regulatory authorities for these diseases. Scientists conducting research have to take special approval to use these therapies such as probiotics and fecal transplantation. Even if you have a lifestyle disease such as diabetes, there is no benefit of knowing your gut-flora composition as yet. Maybe in future, you would walk in with your stools and walk out with a prescription advising a particular bacterium or a cocktail of bacteria. You would consume them, and everything would be fine. Right now, the only time to get your gut microbiome analyzed is if you participate in research studying its role in health or disease.

SECTION 3
THE PURSUIT OF HEALTHY GUT-FLORA

Prevention is better than cure

Chapter 6

IMPROVING GUT-FLORA: THE DO'S AND DON'TS

Warren Buffet says, "Don't put all your eggs in one basket". What he simply means is that there should be 'Diversification' in investment. Those of you who are financially enlightened know about the benefits of diversification. Just like the world of stocks and investments, diversity is also the key to achieving and maintaining a healthy gut-flora. Healthy gut-flora, just like a healthy investment portfolio, is diverse. Persistence is another financial principle that would work wonders in your pursuit of a healthy gut microbiome. For you to enjoy a healthy gut-flora, persistence is non-negotiable. We know that the gut-flora is *resilient;* it always tries to come back to its original state. So, if you wish to change your gut-flora for the better, you will need to work on it continuously otherwise, you might end up losing the gains you achieved.

Now let us move from Warren Buffet to another man, J L Collins. This gentleman is the author of a best-selling book 'The Simple Path to Wealth'. As the name suggests, the book advocates keeping things simple while guiding one towards the pursuit of financial independence. Simplicity is the most under-rated virtue. If you keep things simple in whatever you do in your life, you can be persistent. While pursuing a healthy gut-flora, the simple path is as effective as it is to the wealth. Now, let us explore a simple approach to optimize gut-flora.

What Can We Do to Improve Our Gut-flora?

The gut-flora composition, as we learned is influenced by modifiable and non-modifiable factors. We can obviously change the modifiable ones so, let us concentrate on them. These include diet, exercise, sleep, stress, and medications. Amongst them, diet is certainly the most important as it has the most significant impact on the gut-flora.

Diet

"You are what you eat"

This is what all of us were invariably told, specially as a child when we ran around the house throwing tantrums and being painfully picky about what we wanted to eat and what we didn't!

And most of us have heard the stories from our parents about how in the 'fairyland' the children finished their homework so fast that they had all the time in the world to play their hearts out! They don't even get tired on a sunny summer afternoon. They play and play and play! And always the secret to the invincibility of the fairyland kids was being non-picky eaters and consuming a varied diet!! Apparently, these kids had their leafy greens, pulses, fruits, whole grains, and milk, and stayed away from burgers, pizzas, ice-cream, chips, candies, cola-drinks and fries". They also enjoyed home-cooked meals, and junk food wasn't a part of their life at all. Well! These were the childhood stories. But let us reframe the adage "your gut bacteria are what you eat, and YOU are what your gut-flora is! Seemingly twisted but very true!!

Before we learn how important it is to have a balanced, well-rounded diet to enjoy optimum gut health, let us investigate the opposite; what happens to our gut microbiome in response to consuming fad diets? Before we go into the details, let us sort our basics and understand the meaning of the term 'fad diet'.

A fad diet is any diet that becomes popular all of a sudden only to be replaced by yet another fad diet, in the same way the fashion trends popularize and fizzle out! The fad diets are characterized by drastic claims such as rapid weight loss in a short time or guaranteed cure of a long-standing illness with minimal efforts. The recommendations usually deviate dramatically from the standard dietary recommendations issued by the research organizations and government bodies that formulate the nutritional guidelines based on scientific research. These fad diets are highly restrictive in nature and nutritionally inadequate, with the potential to lead to an adverse impact on health, and in extreme cases, even fatality.

Here are some of the common characteristics of fad diets:

- Supported by no to low scientific evidence.
- Claiming that it is one and only one solution to your concern.
- Highly restrictive, eliminating an entire food group or sometimes allowing only a particular food group.
- Characterized by grandiose claims that often sound too good to be true.
- Claiming that dietary recommendations be tailor-maid based on blood group.
- Claiming that the diet "detoxifies" or 'burns fat'.
- Overly simplified conclusions drawn from complex scientific studies.
- Claiming that sustained, long-term lifestyle modifications, physical activity, and exercise are not required for weight loss and health.
- Endorsed by celebrities.

Let us take a minute now and think about the popular fad diets today that match these characteristics. How many fad diets can you think of? Make a list and ask your friends and family as well. Compare notes for some 'sciencey' fun!

History of Fad Diets

The first accounts of fad diets date back to the middle of the 19th century. This was when the Industrialization and commodification of food supplies started. With the availability of food surpassing the boundaries of 'local', 'regional', 'seasonal' and 'cultural', the adherence with the traditional way of eating began diminishing, which paved the way to fad dieting. One of the earliest fad diets was the Vinegar and water diet which was devised by Lord Byron in the 1820s. However, the first fad diet that got immensely popular is the Banting diet, which was introduced in 1963.

Popular Fad Diets

Though most of us are only aware of the most popular modern-day fad diets such as ketogenic diet and intermittent fasting, the number of fad diets that exist is enormous! If we start describing, we will end up having a huge list. Really!! Let us familiarize ourselves with the names of a few popular fad diets and then discuss the most popular ones in detail, focusing on their key features, merits, and disadvantages.

Some of the fad diets that have enjoyed popularity at some point in time/ continue to be popular include:

- Cabbage soup diet
- Grapefruit diet
- Atkins diet
- "Keto" or ketogenic diet
- Cambridge Diet
- Slim-Fast
- Intermittent fasting
- Juice fasting
- Banting diet

Illustration 4.

Now let us examine the popular fad diets and critically analyze their merits and demerits.

Cabbage Soup Diet

The cabbage soup diet is a drastic, short-term weight loss diet based on consuming cabbage soup as the primary food source for a week to achieve accelerated weight loss. It involves eating practically nothing but cabbage soup in large quantities. The proponents of the cabbage-soup diet claim that this diet can result in weight loss of up to 10 pounds (4.5 kg) in seven days. This diet gained popularity in the 1980s as a fax lore. Although, this diet does result in weight loss because of being extremely low in calories, the weight loss is temporary and regained soon after one resumes the prediet eating patterns. Also, as cabbage soup is very high in fiber, those who follow the diet report flatulence and cramping as the common side effects.

Banting Diet

Banting diet is the first fad diet to have gained remarkable popularity. It was first popularized in 1963 by William Banting, a funeral director based in London. Banting diet is practically the parent diet to all the Low Carb High Fat (LCHF) diets that are popular today. To manage his obesity and other health problems, Banting undertook the dietary changes upon. William Harvey's recommendations, a renowned physician based in London. He benefitted from the dietary interventions by Dr. William Harvey and published them in the form of an open letter. He drafted his diet plan as a personal testimonial in the form of a pamphlet that was popular for decades. William Banting was morbidly obese and also suffered from hearing loss. As per his account, after being on this diet, he regained his hearing and improved his vision.

The Banting diet became the talk of the town again in the year 2014 when Prof. Timothy Noakes popularized Banting Diet in South Africa. A professor in the Division of Exercise Science and Sports Medicine, University of Cape Town, Dr. Noakes named his high-fat, low-carbohydrate diet after Banting. In an attempt to limit cereal intake, people replaced cauliflower with wheat and rice for pizza base and toppings, which, as per the media reports, resulted in National cauliflower shortage in South Africa. Many restaurants started "Banting menus." The Banting diet had become so popular that up until the mid-twentieth century, the name 'Banting' became synonymous with dieting: the verb 'to bant' – as in 'to diet' – appeared in the *Oxford English Dictionary* until 1963.

Intermittent Fasting

Intermittent fasting is one of the most famous modern-day fad diets. It is an umbrella term used to describe different variations of meal timing schedules that oscillate between periods of voluntary fasting and feasting. Intermittent fasting has become a health trend in recent times. The proponents of intermittent fasting claim that it causes weight loss, helps prevent lifestyle diseases and increases longevity.

Here are some of the claimed health benefits:

- Weight loss
- Lower blood pressure
- Improved insulin sensitivity
- Reduced risk of prostate cancer
- Reduction in LDL, triglycerides levels
- Improved memory and attention
- Reduces Alzheimer's and Parkinson's risk

The key variants of intermittent-fasting include:

1. Alternate-day diet
2. Time-restricted feeding
3. 5:2 diet

I. Alternate day diet:

This variation of intermittent fasting involves a modified form of fasting every alternate day. One day is the regular feasting day, while the nest day is the fasting day, on which one is supposed to limit the calorie intake to 500 calories or approximately 25% of the regular daily calorie intake. The feasting and fasting days alternate with each other. There is flexibility about what a person can consume on the feasting days.

II. Time-restricted feeding

Time-restricted feeding allows eating only for a restricted and fixed number of hours every day. While following a time-restricted subtype of intermittent fasting, feeding is permitted only during a specific time window beyond which there is strict fasting. The most common variant of time-restricted feeding is the 16:8 diet that involves 16 hours of fasting a day with an 8-hour window during which eating is allowed.

III. 5:2 diet

The 5:2 subtype of intermittent fasting was popularized by the BBC broadcast journalist Michael Mosley, who experimented with different intermittent fasting variants to address his health concerns. In this subtype of intermittent fasting, five days a week of regular eating is recommended, while for the other two days in the week, a strict calorie restriction below 500–600 per day is recommended.

Limitations/Challenges/dangers unique to intermittent fasting:

1. During the fasting hours/days, one might feel irritable, unable to focus on work, lethargic, and low on energy and productivity due to an inability to replenish the body's calorie and nutrient resources.
2. Forbidden fruit is always attractive. During the non-fasting days/feasting hours, there is a tendency to binge on unhealthy foods that are processed, high in calories, high in trans-fats, sodium, refined sugar, and preservatives that may take a significant toll on one's health.
3. Gastric acidity is commonly observed in response to prolonged fasting.
4. The scientific evidence supporting the claims put forth by the proponents of intermittent fasting is weak.

Ketogenic Diet

The ketogenic diet is also one of the most popular modern-day fad diets. Its popularity can be understood by the fact that 'What is keto" was the second most frequently googled health-related question in the year 2019. Dr. Russel Wilder, MD, Mayo Clinic, first described the ketogenic diet, in the year 1921 as a therapeutic diet for the treatment of epilepsy and later it gained popularity for being effective with weight loss and other health concerns. A ketogenic diet is an LCHF diet; that is, it is Low Carb High Fat. The rationale behind the effectiveness of the ketogenic diet in weight loss is that When one consumes less than 50 grams of carbohydrates in a day as recommended by the proponents of the keto diet, the body eventually runs out of its preferred energy source, the blood sugar, and is instead forced to metabolize fats at an accelerated pace within 3-4 days, resulting in weight loss. This metabolic shift from utilization of carbohydrates to utilization of fats is termed as 'ketosis'.

Proclaimed benefits of the ketogenic diet:

- Prevention and control of heart disease,
- Prevention and management of cancer.
- Prevention of Alzheimer's and Parkinson's disease,
- Management of epilepsy,
- Reversal of PCOD
- Reduction in acne.

However, there are some unique challenges with the ketogenic diet:

1. Keto flu: Upon starting with the ketogenic diet, most people experience keto flu for the initial few days, characterized by fatigue, lethargy, vomit, and gastrointestinal distress. Some also experience bad breath.
2. Because of the restrictive nature of the ketogenic diet, it may result in many vitamin and mineral deficiencies. In the pursuit of restricting the daily carbohydrate intake below 50 grams, whole grains and fruits have to be significantly cut down, depriving the body of dietary fiber, polyphenols, and other nutrients.
3. As the keto diet is a high-fat diet, there is always a risk of excessive intake of saturated fats. This may result in the development of cardiovascular diseases.

The Dangers of Fad Diets

Fad diets have many dangerous effects associated with them. There are reports of people having lost their lives to fad dieting. Different fad diets have some common adverse impacts on health and some that are unique to that particular diet. Let us discuss a few common dangers of fad diets.

1. Nutritional Deficiencies

Specific fad diets involve eliminating entire food groups such as dairy and cereal or severely restricting specific macronutrients such as carbohydrates and fats, resulting in nutritional deficiencies and ensuing health problems. All the macronutrients and micronutrients are required in the diet in the right quantities to optimize health and each one plays a critical role in keeping us free of disease. Therefore, restricting the intake of different food groups/macronutrients translates to jeopardizing our health and well-being and impairing normal body functions. A cabbage soup diet by allowing only the intake of cabbage can lead to various nutritional deficiencies.

2. Lowering of Basal Metabolic Rate (BMR)

By taking up fad diets, we tend to slow down our metabolism. By severely restricting calories while on the fad diet and then resuming our pre-fad diet eating patterns, we end up messing up our metabolism. Such drastic patterns of dietary intake tend to meddle with the body's natural response to hunger and satiety. A slower metabolic rate makes it harder to lose weight.

3. Adverse Health

Because of cutting down on calories drastically and eliminating food groups, one may experience symptoms as diverse as dehydration, depression, acne, fatigue, irritability, nausea, dizziness, inability to focus, irregular bowel movements, constipation, and reeling episodes. Similarly, in the keto diet, because of extremely low carbohydrate intake, the breakdown of proteins is expedited resulting in muscle wasting. This indirectly impacts the amount of physical activity and exercise performed as because of the aforementioned issues; one doesn't feel motivated/ energetic enough for physical activity.

4. Social Isolation

Food has significance in our life way beyond nutrition. Food is associated with festivities and celebrations, and preparing food and eating it together is a powerful medium for the expression of love. Limiting food choices in pursuit of fad diets and consequently following meal plans that are restrictive and prohibit one to eat from the family pot can result in feelings of isolation. Before taking up any dietary regimen, we need to ask ourselves: "Can I follow this diet for the rest of my life?" If the answer is no, it is wise to say no to the fad diet in question!

5. Disordered Eating Patterns

When the dietary restrictions dictate which food group, at what time, and what quantity we are allowed to eat, it disturbs our relationship with food. This might, in some cases, culminate in the development of disordered eating patterns. When on a highly restrictive diet, people often dwindle between the periods of extreme calorie restriction, followed by relentless binging. This phenomenon is also referred to as the 'dieting pendulum' crisply representing the phenomenon of constant swings from phases of extreme restriction to binging, leading to psychological distress and chaos.

6. Poor Self-Esteem

Because of the overly restrictive nature of the fad diets, they are hard to follow. More often than not, people find themselves unable to adhere to these diets beyond a few weeks or a few months. This takes a toll on their self-esteem as they perceive themselves as 'failures' due to the inability to stick with the fad dietary regimen.

Why Do We Keep Falling for Fad Diets?

The popularity of fad diets is at an all-time high today, and just about every few days, we witness a new diet craze. These fad diets profess drastic dietary eliminations and vary from low-fat, to low-carb, to blood group-based diets. Some retain their popularity for months and years, while the most have a short life span. The question arises- why are fad diets so popular after all? Despite knowing that fad diets can have disastrous consequences on our health, why do we continue to fall for them? What is that element in fad diets that attracts us to such a degree that we go to the extent of eliminating entire food groups from our diet? What isn't the scientifically sound advice - Eat right and move more gripping enough compared with the claims put forth by the proponents of fad diets? Why is it that we trust the claims associated with fad diets despite the lack of concrete scientific evidence in their favor? To sum it all up, why do we embrace fad diets? Let us explore the answer.

1. Time Frame

Fad diets promise big results in short time contrary to the scientific recommendations that emphasize a life-long commitment to eating well-rounded nutritious meals encompassing all food groups- fruits, vegetables, dairy, cereal, and protein. Exercise and physical activity are an integral part of these recommendations. The fad diets come with the promise of drastic results in months or even weeks. This is a psychological trap that attracts one towards the 'easier way out', which is committing to a dietary pattern, how so ever restricting for a short time, instead of a lifetime commitment to adopting a healthy lifestyle. Quick fixes are attractive.

2. Feeding on the Vanity

In contrast with lifestyle changes, fad diets focus more on the weight loss and inch-loss compared to gaining health and preventing or managing diabetes, high blood pressure, heart diseases, etc. Adolescents and young adults tend to place greater emphasis on appearance than health, and this is one of the significant reasons why fad diets have a strong appeal for them.

They feel more motivated to change the way they look compared to changing their state of health or working on the prevention of potential lifestyle diseases. For the young and impressionable minds, slimming down for a wedding may feel like a bigger goal than reducing the risk of developing diabetes in the future.

3. Grandiose Claims

While most scientific advice is straight forward and worded with the intention to educate and guide, the claims made by proponents of fad diets are loaded with grandiosity. The purpose is to portray the diet in a way so that it seems like the one and only solution for the concern at hand. A strong illusion of scientific credibility is also created.

4. The Magical Element

We, as human beings, love magic. The thought that this single element can help me lose weight or this single food group that was the cause of all my troubles or this miracle berry that is going to do the trick for me is exciting. Something as straightforward as 'eat well and work out well' for a lifetime isn't so exciting after all!

5. Celebrity and Media Endorsements

The young and impressionable tend to emulate celebrities. If their favorite celebrity endorses a particular fad diet, they tend to follow it too. Also, repeatedly watching the media advertisements about various fad diets positions them in our minds firmly.

6. Peer Pressure

If everyone around us is following a particular fad diet, we tend to get influenced too. We, as human beings, tend to emulate the feeding behaviors of those around us. When a fad diet becomes popular, and we see our friends and family members following it, we feel encouraged to follow it too.

Gut Microbiome and Fad Diets

We have discussed the impact of fad diets in influencing various aspects of our well-being, both physical as well as psychological. Let us now understand the effects of fad diets on gut microbiome structure and function. Let us look at the keto diet, one of the most popular modern-day fad diets, and its effect on the gut-flora.

As we have learned before, the ketogenic diet is a low carbohydrate, high fat (LCHF) diet. With that knowledge in mind, let us understand how the consumption of a diet high in fats or low in carbohydrates or a combination of both impacts the gut health. Wit et al. investigated the impact of a high-fat diet on gut bacterial diversity and composition. They observed that a diet high in saturated fats resulted in a decrease in the diversity of gut bacteria and increased the ratio of Firmicutes-to-Bacteroidetes (Wit et al. 2012). The study results suggest that the intestinal flora composition may be adversely affected by the intake of a high-fat diet such as ketogenic diet. Another study also reported a reduction in the density of Bifidobacterium in response to the consumption of a high-fat diet for 25 weeks (Zhang *et al.* 2010). As you already know, Bifidobacterium has a protective effect on the gut epithelium. He et al. investigated the effects of the consumption of a carbohydrate-restricted diet on the composition of intestinal bacteria in the mice models. A decrease in the density of Enterobacteria that exert an anti-inflammatory effect on the host was observed with a concomitant increase in the density of inflammation-inducing Enterobacteria upon consumption of a carbohydrate restricted diet (He et al. 2020). A particular class of carbohydrates called the Microbiota accessible Carbohydrates (MACs) is critical for the gut bacteria to thrive, as discussed previously. Since the ketogenic diet is a low carbohydrate diet, recommending an intake of fewer than 50 grams of carbohydrates a day, there is only a limited intake of whole grains and fruits, two essential sources of MACs. A diet low in MACs is associated with microbial dysbiosis, which results in a host of health problems in the host (Sonnenburg & Sonnenburg 2014). Thus, ketogenic diets characterized by low intake of carbohydrates are deficient in MACs that affect both the diversity of gut bacteria and the density of beneficial bacteria in the gut.

A few scientific studies have also looked into the effect of intermittent fasting on the gut microbiome. In a mice study, alternate day intermittent fasting increased the number of Firmicutes, leading to increased production of short-chain fatty acids acetate and lactate in the intestines providing health benefits (Li et al. 2017). Another study that analyzed the gut-flora before and after 29 days of Ramadan fasting showed an increase in the *Akkermansia muciniphila* (a good friendly gut bacterium) after the period of fasting (Özkul et al. 2019). Ramadan fasting is from sunrise to sunset during the holy month of Ramadan. People eat a large meal after sunset and a lighter meal before sunrise. The fasting period is approximately 12 hours; however, this period may vary depending on the geographical location. The beneficial effects of intermittent fasting on gut-flora are probably because it does not restrict the diet composition. It instead prescribes time restrictions. In a study published in the highly reputed Journal of the American Medical Association (JAMA), unrestricted daily eating was as effective as 16-hour fasting intervals for weight loss. Both the groups had similar calorie intake (Lowe et al. 2020). Although, as per the latest scientific evidence, intermittent fasting is rather beneficial for gut bacteria, it is not particularly useful for weight loss as compared to a balanced diet consumed without time restrictions.

To conclude, fad dieting is harmful from the perspective of gut health and overall well-being.

Now you may ask, what is the right approach to eating? How should our plate look like?

The answer is simple and straightforward. The scientific organizations that formulate dietary recommendations throughout the world make similar nutritional recommendations. If we look at the nutritional recommendations by the US Department of Agriculture (USDA), USA, and compare them with the recommendations by the National Institute of Nutrition (NIN), India, we will find striking similarities. Essentially, both recommend a similar pattern of eating.

MyPlate, the nutrition guide published by the US Department of Agriculture, Centre for Nutrition Policy and Promotion, shares ten important tips to guide US citizens to eat healthily:

1. *Find your healthy eating style*
2. *Make half your plate fruits and vegetables*
3. *Focus on whole fruits*
4. *Vary your veggies*
5. *Make half your grains whole grains*
6. *Move to low-fat or fat-free milk or yogurt*
7. *Vary your protein routine*
8. *Drink and eat beverages and food with less sodium, saturated fat, and added sugars*
9. *Drink water instead of sugary drinks*
10. *Everything you eat and drink matters*

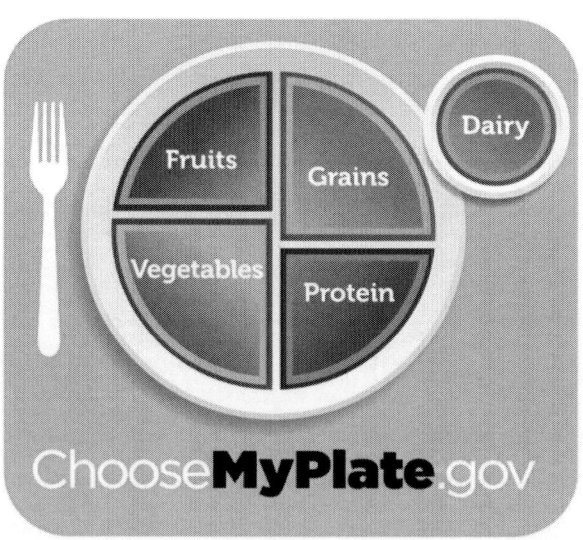

Similarly, the National Institute of Nutrition, India, recommends the following:

- *Choose a variety of foods in amounts appropriate for age, gender, physiological status, and physical activity.*
- *Use a combination of whole grains, grams, and greens. Include jaggery or sugar and cooking oils to bridge the calorie or energy gap.*
- *Prefer fresh, locally available vegetables and fruits in plenty.*
- *Include in the diets foods of animal origin such as milk, eggs, and meat, particularly for pregnant and lactating women and children.*
- *Adults should choose low-fat, protein-rich foods such as lean meat, fish, pulses, and low-fat milk.*
- *Develop healthy eating habits and exercise regularly, and move as much as you can to avoid a sedentary lifestyle.*

FERMENTED FOODS

The recommendations by various nutrition societies, as we just learned, emphasize a well-rounded balanced diet but they also miss one component vital for a healthy gut-flora. While a diverse, balanced diet allows beneficial bacteria to grow in our intestines, what if we can simply eat good bacteria and enjoy their benefits? Fermented foods will enable us to achieve that exactly. The fermented foods are rich in good friendly bacteria, also known as probiotics. They improve the gut-flora by simply providing healthy bacteria to our intestines. Different sources of probiotics are popular in different parts of the world. Some of the common examples of probiotics include- Curd, yogurt, Kefir, Miso, Sauerkraut, and Kimchi. Including fermented foods, in one meal every day can be highly beneficial for the gut-flora. Even scientific studies have shown the benefits of fermented foods on the gut microbiome. The widely popular Korean fermented food Kimchi, is proven to improve the gut-flora composition to a healthier state (Jung et al. 2011). Fermented foods provide a range of health benefits. They have been shown to improve blood pressure (Kim et al. 2008), reduce bad cholesterol levels (Choi et al. 2013), and prevent breast cancer (Yamamoto et al. 2003) in various scientific studies. It is imperative to include one of these fermented foods in your diet. You can choose depending on where you live and which fermented food is available based on your local food tradition.

EXERCISE

Exercise, as you know, is another modifiable factor that has an enormous impact on gut-flora. If you want a healthy gut-flora, you must be regular with your runs or swims or gym sessions, whatever you love to do. Although, as we learned in chapter 1, exercise begins to improve the gut bacteria composition within six weeks, we do not yet have an established guide as to what type of exercise one should practice and how much should be the daily duration to improve the gut-flora. However, we can safely follow the recommendations by various societies such as the American college of cardiology for healthy adults (Piercy et al. 2018). These guidelines recommend:

- At least 150 minutes of moderate aerobic activity or 75 minutes of vigorous aerobic activity every week.
  and
- Strength training exercises on two or more days a week that work all the major muscles.

Gut-flora diversity is shown to increase in relation to improved aerobic fitness. But to derive maximum health benefits, one should practice a mix of both aerobic and strength exercises. If you already exercise, it will keep your gut-flora healthy besides providing several other benefits; if not, you have one more reason to start.

Sleep

Most of us feel proud of our clean diet and pat ourselves for adhering to our exercise schedules but still do not give the same importance to sleep. All of us have boasted about the hours of sleep (or the lack of) we can thrive on. We are impressed when our colleague tells us that he/she sleeps only 3-4 hours in a day and even long for that efficiency secretly. But the sleep is irreplaceable. It is paramount not just for our mental but also physical well-being. Lack of sleep can lead to obesity and related illnesses. Even most of the repair work in our body is done while we are asleep. If you are regular in your gym but not in your sleeping habits, you will not build muscles as the gain in muscle mass occurs during repair. Exercise in itself is a catabolic process and leads to the breakdown of muscles, which grow as they are repaired during rest and sleep.

Like exercise, sleep impacts gut-flora, but we do not know about the exact duration of sleep we must get to achieve a healthy gut-flora. Here too, like exercise, we can follow the general recommendations for healthy adults. Most of the scientific guidelines recommend having at least 7 hours of uninterrupted sleep every night for a healthy adult. Sleeping less than 7 hours per night regularly is associated with adverse health outcomes, including weight gain, obesity, diabetes, hypertension, heart disease and stroke, depression, and increased risk of death. It is also associated with impaired immune function, impaired performance, and a greater risk of accidents (Watson et al. 2015). If you wish to keep your gut-flora healthy, you must sleep for more than 7 hours every night regularly. Isn't it amazing to know that you can simply sleep your way to the healthy gut-flora?

Stress Management/Relaxation Techniques

The next crucial modifiable determinant of gut-flora composition is stress. We know that stress is a normal physiological response that is essential to protect humans from threats. However, uncontrolled and chronic stress has a damaging effect on overall health and well-being. In the last decade, there has been a global increase in awareness about the importance of psychological well-being and the need to manage stress effectively. More and more people are taking up relaxation techniques such as meditation to manage stress and anxiety. Meditation is becoming popular even as an adjunct therapy in the management of anxiety and other psychological disorders. If stress impacts gut-flora, then does the relaxation techniques used to manage stress improve the gut-flora? Let us again look at the scientific evidence. Househam et al. reviewed the studies investigating the role of meditation in favorably altering the gut microbiome in a way that enhances the overall well-being. The authors searched multiple reputed scientific databases with the relevant keywords to investigate the link between stress management techniques and alternations in the gut microbiome. They concluded that meditation helped the volunteers/subjects regulate the stress response, thereby significantly lowering chronic inflammation and enhancing the integrity of gut-barrier function (Househam et al. 2017). Meditation was found to mitigate the effects of psychological stress that throws the body into the fight-or-flight response, which leads to increased production of catecholamines and corticosteroids. Relaxation techniques provide a wide range of health benefits ranging from anxiety disorders to irritable bowel syndrome and joint pains. They are a must-do in your journey towards a healthy gut-flora.

Alcohol and Smoking

Both alcohol consumption and smoking have an adverse impact on gut-flora. In an elegant scientific study, Lin et al. enrolled 116 healthy males divided into four groups:

Group A: Men in group A neither smoked nor consumed alcohol
Group B: Included men who smoke but didn't consume alcohol
Group C: Included men who didn't smoke but consumed alcohol
Group D: Included men who smoke and consumed alcohol

They collected the fecal samples from all four groups and assessed them for gut microbiome composition and the Short-chain fatty acid (SCFAs) levels. Gut microbiome composition was found to be drastically altered in response to alcohol consumption and smoking. SCFA levels were lower in groups that consumed alcohol and smoked. Group B, C, and D had a higher concentration of Bacteroides, which correlated with the number of cigarette packs consumed per year. Group D and group B had a similar gut-flora composition. Still, it was different from group C. The study results indicate that cigarette smoking and alcohol consumption impact the gut-flora differently, with smoking playing a more significant role in gut dysbiosis than alcohol consumption (Lin et al. 2020). Smoking cessation has also been shown to change the gut-flora composition. In a study of 10 healthy subjects undergoing smoking cessation, it led to an increase in the diversity of gut-flora as well as an increase in Firmicutes and reduction in Proteobacteria (Biedermann et al. 2013). If you don't smoke and don't habitually drink alcohol, it will be easier for you to achieve healthy gut-flora. And if you do, then stopping will undoubtedly improve it. So please attempt to stop.

Antibiotics

We have already learned that antibiotics don't differentiate between friendly gut bacteria and harmful pathogens. But they are life-saving medicines and must be used judiciously. You just need to follow a simple principle to protect your gut-flora from the detrimental effects of antibiotics. Don't self-medicate. Take antibiotics only when prescribed by a physician.

Probiotic Supplements

You already know that probiotics are live bacteria, which when consumed, provide health benefits. One way to enjoy their goodness is by including fermented foods in your diet, which are a rich source of these bacteria. Another is the probiotic supplements, which contain these beneficial bacteria. There are a wide variety of probiotic supplements available. You go to any nearby supermarket; you will find a big section devoted to them. You google them, hundreds of search results are available in less than a second. How to choose which one suits your needs? Do you really need them? Can you afford them for the long-term as most of them are quite expensive? You know that gut-flora is resilient, and any change you wish to incorporate to make it healthy will need to be permanent; probiotics being no exception. Let us try to answer these.

You may recall that although the bacteria reside throughout our gut, the large intestine has their maximum concentration; 10^9--10^{12} or 1 billion to 1000 billion live bacteria per ml of its contents. Along with the huge number, we also have a large variety of bacteria in the gut-flora, around 500-100 different species. Just look at the number of bacteria present in the probiotic supplements, and you will find that there is a wide range reaching up to billions per supplement dose. The probiotic supplements are available as a varied range of products such as tablets, capsules, sachets, and liquid formulations. Even when the number of bacteria is in billions, it equals the amount present in just 1 ml of the contents of your colon or large intestine. We know these are the 'good healthy bacteria,' but most of the bacteria routinely present in the gut are also good and healthy. Let us understand with an example. The probiotic VSL#3[3] is one of the most used and prescribed probiotic, and amongst the few which have been shown to be effective as a treatment for any illness. It contains eight different strains of probiotic bacteria, including four strains of Lactobacilli, three strains of Bifidobacteria and *Streptococcus thermophiles*; is available in packets and capsules and contains 900 billion live bacteria in the packet which has the highest dose, the number present in 1 ml of colonic contents.

If VSL#3 contains 900 billion live bacteria, then why all probiotic supplements do not contain these much or more? Simply because providing the live probiotic bacteria in a capsule form is expensive. Another fact you must know is that most of the regulatory authorities,

[3] VSL#3 is now sold as visibiome in the markets worldwide.

including the FDA, categorize them as dietary supplements and not as medicines that are subject to stricter controls. Even VSL#3 is classified as 'refrigerated medical food'! Only god knows what it means. You can look on your own at the FDA website. Medical food can be used only under medical supervision. This categorization as supplements rather than medicines allows for a wide range and different types of bacteria in the probiotic supplements. If you compare the leading brands in the markets, most of them contain 5-20 billion live bacteria per serving. Similarly, both single strain probiotics and brands with a cocktail of bacteria like VSL#3 are available. Another big challenge with probiotics is their fragile nature. They are live bacteria and must survive the process of manufacturing, supplies, and storage. Besides, they can be killed by the acid present in the stomach. Even after your purchase, they must be kept under prescribed conditions. It is wise always to read the labels and store them as the label says.

Now let us look at the benefits offered by the probiotic supplement. You have already read that gut-flora have a role in the development of lifestyle diseases. Probiotics are being evaluated as a prospective treatment for them in various scientific studies. You might recall that probiotic supplements have shown some benefits in diabetes, obesity, and autism spectrum disorders, but as yet, they are not established as a treatment for these illnesses. The only diseases in which probiotics are medically prescribed and have a proven role are irritable bowel syndrome (Ford et al. 2014), ileal pouch inflammation (pouchitis) (Nguyen et al. 2019), and prevention of antibiotic-associated diarrhea (Goldenberg et al. 2019).

So finally, which is one to choose? If you have a medical condition such as IBS or ileal pouch inflammation, you must take only the approved probiotic supplements under medical supervision. Many of the treatment benefits of probiotics are dependent on a particular strain or a specific combination. Self-medication can be harmful as it may delay prescribing the other, more effective treatment options. Let the physicians decide which one you need and whether you need a probiotic or a medicine. If you wish to take them for health benefits, we would suggest incorporating more fermented foods in your diet according to your tradition and local food practices. But if you must take a probiotic supplement, it is wiser to choose the one with more strains and a larger number of bacteria.

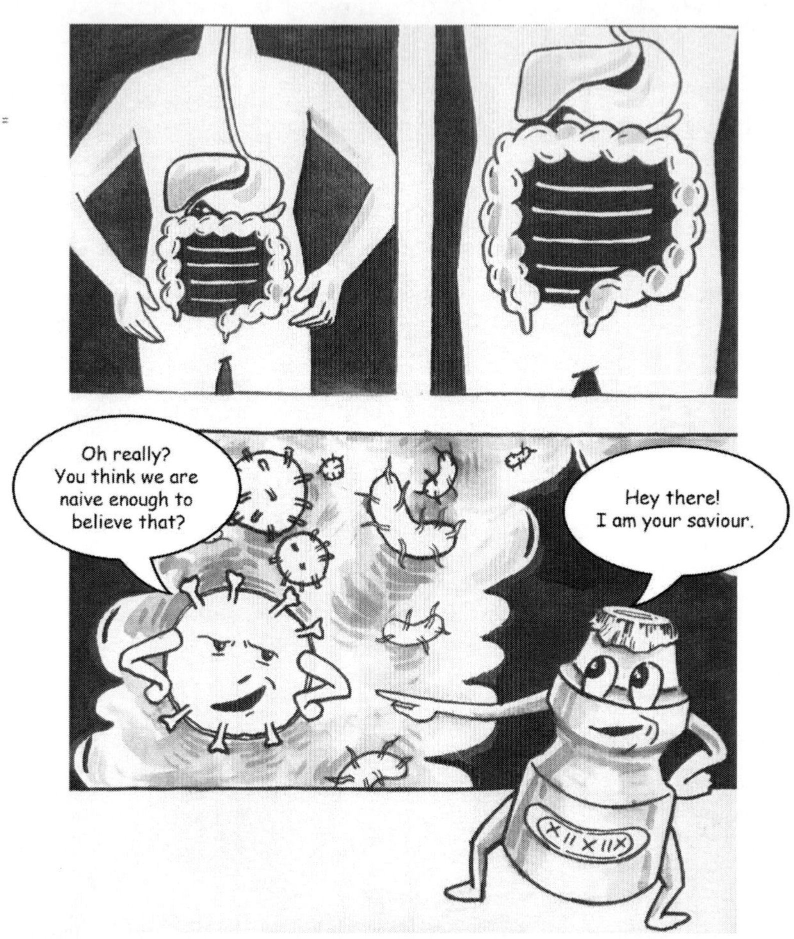

Illustration 5.

Do you need to take one? Well! If your physician prescribed for a particular medical condition, definitely take it. For general health benefits, you could get the good bacteria from the fermented foods themselves.

Do you need to take long-term? You want a permanent change in gut-flora, adopt the healthy practice permanently. Studies have shown that the beneficial changes in gut-flora induced by probiotic supplements take one month to develop and wean off once they are no longer consumed. You must take a probiotic supplement at least for four weeks before you expect a change in gut-flora. And Yes! If you have indeed zeroed-in on the supplements for your regular dosage of healthy bacteria, you need to permanently stick to them. However, it may be more feasible and cheaper, in the long run, to instead consume fermented foods.

FECAL MICROBIOTA TRANSPLANT (FMT)

When we are keen to achieve a healthy gut-flora, what better than getting a new one from a supremely healthy donor? If diseased hearts and livers can be replaced, why can't the disease-causing gut bacteria can't be replaced with vibrant, friendly ones? Well, well, point taken!! Microbiota transplant is here.

This microbial transfer of good gut bacteria is achieved by transplanting the feces from the donor to the recipient, thus termed as fecal microbiota transplant (FMT). You could alter the gut-flora by taking antibiotics or probiotics, but FMT is altogether a different ball game. It provides a much larger quantum of gut-bacteria to the recipient. The process of FMT is a complex one. It starts with identifying a suitable donor, screening him for various diseases which can be transmitted through the fecal matter, preparing the stools for the transplant, and finally delivering it to the recipient.

None of us ever thought that one could be a millionaire just by collecting the poop of another millionaire. But the day is not far away. Soon you could be seeing advertisements asking for healthy fecal donors. How would an ideal fecal donor look like? We can only speculate. S/he would be someone free from lifestyle diseases, someone who consumes a rich diverse diet, lean, physically active and free from addictions such as smoking and alcohol; and with a large number of good friendly gut bacteria. Now you are thinking about the friend who could be the next millionaire! Isn't it? Some scientific trials recruit the stool donors by advertising, while a healthy relative is also a reasonable donor. In fact, for recurrent *Clostridium difficile* infection, the disease for which FMT is established as the first-line treatment, a healthy relative who himself is free of disease is a perfectly acceptable donor. But in future, if FMT becomes the go-to therapy for obesity, diabetes, stress-related, and behavioral disorders, then we shall need special donors and stool banks. Even one's own stools could also be used for FMT for diseases such as inflammatory bowel diseases, which have disease-free periods in between. Stools collected during disease-free periods can be used during the relapse (Duplessis et al. 2012).

A stool donor is rigorously screened. They undergo an exhaustive preliminary interview, which is followed by labs. During the interview, a donor is asked about the usage of antibiotics, chemotherapy, immunosuppressant drugs, and the long-term proton pump inhibitors over the last three months. He/she tells about the history of any chronic gastrointestinal disease, gastrointestinal cancers, autoimmune disorders, known blood-borne infections such as hepatitis B, hepatitis C, HIV, HTLV I and II, obesity, and neurobehavioral conditions. Finally, they also need to inform about high-risk behaviors, illicit drug use, needle stick injury, body piercing, tattooing, acupuncture in the last six months, and recent acute gastroenteritis, and vaccination with any live vaccine (Bibbò et al. 2020). After that, they are subjected to a large number of blood and stool tests (Table 1). FMT is a highly specialized procedure and has inherent safeguards and controls to ensure the recipient's safety. Donors are screened to select the ones with good gut bacteria and who are free from diseases that can be transmitted through the FMT. If you ever need a microbiota transplant, you must get one under medical supervision at specialized centers. Cutting corners may do more harm than benefit.

There are various methods to perform a fecal microbiota transplant. The stools can be made into liquid slurry and delivered by enemas and through a colonoscope straight to the large intestines. Although not aesthetically pleasing, the stools can also be delivered to the gut by upper gastrointestinal routes either through the scope or using the feeding tubes, which are put through the nose into the stomach or jejunum. However, the latest and definitely the most acceptable method (to the recipient) of stool delivery is oral capsules. It involves the delivery of stool mixed with a cryoprotectant, most commonly glycerol, and double- or triple-encapsulated to protect the stool from stomach acidity. Studies have shown the oral capsules to be equally effective to the colonoscope delivered fecal transplants (Kao et al. 2017).

Table 1. Screening tests recommended for stool-donors

Blood investigations: 1. Complete Blood Count 2. Liver function test 3. Serum creatinine 4. C-reactive protein 5. Serology for hepatitis A, hepatitis B, hepatitis C, and HIV viruses 6. Testing for syphilis
Stool investigations: 1. Clostridium difficile infection 2. Routine bacterial culture for enteric pathogens (*Salmonella, Shigella, Campylobacter,* and *Escherichia coli* serotype O157) 3. Parasitic infections such as G*iardia lamblia, Cryptosporidium*, Isospora, and Microsporidia 4. Various other infections such as Norovirus, Rotavirus, H. Pylori, etc., depending on the donor or recipient characteristics. 5. During the covid-19 pandemic, the PCR test for SARS CoV-2 is also recommended.

Abbreviations: HIV, Human immunodeficiency virus; SARS Cov-2, Severe Acute Respiratory Syndrome Coronavirus-II.

And lastly, who needs a microbiota transplant? A glance at *ClinicalTrials.gov,* a database of all the clinical trials including those currently undergoing, shows that more than 90 trials evaluate the role of FMT in a myriad of diseases ranging from liver failure to autistic spectrum disorders (ASD) are in the progress at present. Obviously, many gastrointestinal illnesses are also being studied along with lifestyle diseases such as hypertension, neurological disorders, and various cancers. As already mentioned, FMT is the first-line treatment for recurrent *Clostridium difficile* infection. Besides, it has also shown promising results in inflammatory bowel diseases (Paramsothy et al. 2017) and IBS (Johnsen et al. 2019). Although FMT is being studied as treatment for a large number of diseases, the definite indications as of now are few. Apart from clinical trials, only those suffering from any such disease should undergo FMT.

In a nutshell, the simplest way to achieve a healthy gut-flora is:

1. Consume a rich, diverse, and balanced diet. You can increase the diversity of your existing diet in the following ways:
    a. Include multiple kinds of cereals in your diet every day rather than just keep on eating one or two kinds of cereals.
    b. Consume seasonal fruits and vegetables. It will automatically ensure diversity and would also increase the amount of fiber in the diet.
    c. Avoid processed food. Consume whole grains, which provide complex sugars and fiber that act as prebiotics.
    d. Include fermented foods in at least one meal every day.
2. Exercise regularly, do a mix of both cardio and strength training for maximum benefits.
3. Maintain sleep hygiene to ensure adequate sleep of more than 7 hours every night.
4. Practice a relaxation technique to better deal with stress.
5. Avoid alcohol and smoking.
6. Do not self-prescribe medications in case of any illness, especially antibiotics.
7. Be persistent in your efforts.

Now you know; why do you need a healthy gut-flora and what to do and what not to achieve it? Let us look at HOW to do this. How do you achieve and maintain a rich, healthy, and diverse gut-flora? Let us look into that in the final chapter.

Chapter 7

PRACTICAL TIPS FOR OPTIMIZING GUT HEALTH

Q: I understand that a diverse diet is a key to a healthy gut-flora. How can I make my diet more diverse? I usually consume three meals a day. Is it mandatory for me to consume more meals to increase diversity?

A- You're absolutely right in your understanding that a diverse diet is the key to a healthy gut-flora. Let us explore the strategies to enhance diversity in the diet and, along the way, answer your question about the number of meals a day that are optimum for gut microbiome health.

1. You can start your day by having a fruit. Have a different fruit every day and change according to seasons. Each fruit is unique in terms of its nutrient composition. Previously, in the section about diet, we have discussed the role of polyphenols in enhancing gut health. Different fruits are composed of different types of polyphenols that help gut-flora flourish. Eating a different fruit every day would ensure that your diet is rich in various types of polyphenols.
2. Incorporate nuts and seeds into your diet. Have a mixture of flaxseeds, sunflower seeds, pumpkin seeds, chia seeds, almonds,

walnuts, cashew nuts, and pine nuts as a mid-meal snack, or use the mixture as salad toppings. The nuts and seeds are an excellent source of polyphenols, proteins, and microbiota accessible carbohydrates.

3. Have a different breakfast every day. Set out a meal chart and enlist seven breakfast options for seven days of the week. Make a list of the ingredients before the beginning of every week and carry it along while grocery shopping so that all the ingredients are readily available before breakfast preparation.
4. For lunch and dinner, ensure that you have a different vegetable every day and a different pulse/lentil every day.

   Also, instead of having your preferred cereal every day, incorporate various cereals in your diet. For example, you can have a wheat-based meal in the lunch, and rice-based during dinner. You can also incorporate other cereals such as millets that are of different types such as finger millet, pearl millet, etc. to further increase your diet diversity. One simple way to use multiple kinds of cereals in North Indian households is to use multi-grain flour to make breads. You can use such a mixture based on your regional and traditional diet culture and cuisine.
5. We have provided an extensive list of probiotics as well as prebiotic foods that are critical for enhancing gut-health in this book. Incorporate them into your everyday diet. Consume fermented foods at least during one meal daily.
6. Ensure that you have local, seasonal, and regional foods as much as possible. By following this approach, you will have a diverse diet without even giving it a thought. Eating seasonal equals eating diverse as different fruits, vegetables, and grains grown in different seasons. Eating local implies eating what's in the season and also eating fresh.
7. Incorporate raw foods such as salads and sprouts in the diet as between-meal snacks.
8. Avoid eating processed and packaged food. Knowing what to eat is important but knowing what to avoid is equally important.

Essentially, a varied or diverse diet is a diet that includes elements from all the food groups- Fruits, vegetables, dairy, protein, and cereal, and other different types of foods within a food group. A diverse diet supplies the body with a broad spectrum of vitamins, minerals, macronutrients, polyphenols, prebiotics, and probiotics.

Instead of increasing the number of meals consumed per day, the right approach is to increase diversity in the three main meals and to have 2-3 healthy snacks a day as suggested.

Q: What is the difference between curd and yogurt? Which is better for gut health?

A- The strains of bacteria used to ferment yogurt include but may not be limited to *Lactobacillus bulgaricus* and *Streptococcus thermophilus*. The starter cultures used to manufacture yogurt are standardized, and batch to batch variability is low. In the case of curd, the starter culture is non-standardized and is predominantly composed of lactic acid bacteria. Yogurt is primarily a commercially manufactured product with stabilizing agents, preservatives, and, in some cases, added flavors and colors. At the same time curd is traditionally prepared in Indian households, using the curd itself as a starter culture. As the starting culture of curd is non-standardized, the bacterial composition of curd shows batch to batch variability. It is difficult to choose between curd and yogurt as the winner in terms of enhancing gut health but indeed, the artificially added stabilizers, colors, sweetening, and flavors aren't the best deal for overall health, including gut health.

Q: I am always on the go and can't allocate too much time to cooking. Please suggest to me some options for optimizing my gut health.

A- As we have already understood, our gut health is important not just from the perspective of our digestive health but overall physical as well as mental well-being, making it essential for everyone, irrespective of their lifestyle, to incorporate strategies to enhance gut health. Fortunately, diet is a very important modifiable tool to achieve optimum gut health.

The key idea is to incorporate a combination of prebiotics and probiotics in the diet and ensure that the diet is diverse.

Prebiotics, as discussed, are the non-digestible components of our diet that evade digestion by the human digestive enzymes and act as the food for good bacteria in the gut.

Here are a few quick sources/tips to enhance gut health for everyone irrespective of their lifestyle and availability of time:

1. Whole grains: Avoid refined grains and include whole grains in your diet. A small change like substituting refined flour with whole-grain flour to prepare bread on a regular basis can go a long way in nourishing gut bacteria. It will provide increased quantities of undigested carbohydrates to them and will act as prebiotics.
2. Include flaxseeds in your diet. These can be a wonderful addition to your salad/curd and a healthy snack to munch while on the go!
3. Banana and apples have excellent content of prebiotics. Have them as mid-meal snacks. They are very convenient to eat on the go and very rich in polyphenols, which have antioxidant action. These polyphenols greatly improve gut health while protecting the body from oxidative stress, which leads to aging.
4. Use unpeeled potatoes for cooking. Potato skin is an excellent source of prebiotics.
5. Use honey as a sweetening agent for desserts and beverages instead of using refined sugar. This will reduce the intake of simple carbohydrates. For carbs, the complex is better.
6. Try to incorporate different grains in the diet, which will provide much-needed diversity. Citing the Indian example, traditionally, Indians consumed different grains such as rice, wheat, sorghum (jowar), pearl millet (bajra), finger millet (ragi), etc. based upon the seasons and food traditions associated with different festivities and fasts. However, the modern Indian way of eating has shifted

towards the consumption of wheat and rice predominantly. However, this trend seems to be changing again in the light of the recent research indicating the importance of consumption of different grains on gut microbiome composition and overall health.

You can incorporate all these tips and tricks to enhance your gut health despite being pressed for time.

Q: I have tried everything under the sun to lose weight. Nothing works for me. Does this have anything to do with my gut-flora? What can I do to lose weight?

A- Yes, you're right. The composition of our gut bacteria significantly influences our capacity to lose weight. While the research about the mechanisms that link the composition of gut-flora to our ability to lose weight is still underway, there are many conclusive studies about the differences in characteristics of the gut-flora composition of obese versus non-obese individuals.

The gut bacteria regulate our weight in three fundamental ways:

- By controlling the amount of food we consume. You may recall that short-chain fatty acids, especially butyrate, play a key role in providing the satiety signal to our brains. Butyrate is exclusively produced by gut bacteria.
- By modifying the expression of genes directly or indirectly involved in regulating weight loss.
- By influencing how the energy is utilized from the food we eat.

The ability to lose weight is determined by the two key features of the gut-flora.

1. The diversity of the gut bacteria.
2. The nature of the gut bacteria.

Overall reduced diversity of bacterial species is observed in individuals who find it hard to lose weight.

Indeed, the composition of the gut-flora can be altered with consistent, long-term efforts in the right direction. This book provides a detailed description of the strategies to enhance gut health that the readers can quickly implement.

Q: My family enjoys frozen foods, and I love the convenience that comes with the two-step thawing and frying instead of preparing elaborate meals. Are frozen foods safe from the perspective of gut microbiome health?

A- It is absolutely true that frozen foods have a big advantage in terms of ease of preparation and cost. We often find ourselves pressed to prepare elaborate meals, and frozen foods seem like an obvious choice because of the present-day demanding work-life culture. First, let us understand the meaning of the term frozen foods. For the purpose of this discussion, frozen foods imply the commercially available packaged frozen foods such as pizzas, potato chips, waffles, and other snacks that are kept in a frozen state and then heated/microwaved/fried before eating and not the food that is home cooked and frozen for later use. The frozen foods have a detrimental effect on overall health, including the composition and function of the gut microbiome. The frozen foods have a high concentration of hydrogenated palm oil that is laden with harmful trans-fat. It is also loaded with preservatives, sodium, and high fructose corn syrup, the excessive consumption of which is implicated in many lifestyle diseases, including obesity, diabetes, hypertension, altered lipid profile, heart diseases, thyroid disorders, PCOD, and certain types of cancer.

One must refrain from feeding kids frozen foods as the composition of diet in kids tends to have a dramatic impact on their gut microbiome composition as adults. Kids who are obese tend to be obese as adults. We have already discussed the tips/strategies for optimizing the gut microbiome through dietary interventions for those pressed for time. You may consider incorporating them into your everyday lifestyle for optimizing gut health instead of consuming processed food.

Q: I have a history of antibiotic use, and because of a lack of awareness, I didn't follow through with the prescribed courses. How can I restore the good bacteria in my gut?

A- We can understand your concern. Indiscriminate and unstructured use of antibiotics is damaging not just for gut health but for overall health and well-being. It results in antibiotic resistance, which is a global public health concern. Global health organizations such as the Centre for Disease Control and Prevention (CDC) and the World Health Organization (WHO) emphasize that antibiotic resistance is a worldwide crisis. There is an enormous volume of scientific studies that have been conducted across the globe to examine the impact of antibiotic use on human gut microbiome composition. The consensus is that chronic/inappropriate antibiotic use causes a remarkable reduction in the gut bacteria diversity. The oral use of antibiotics may result in the eradication of beneficial species of bacteria from the gut and result in a relative increase in the proportion of pathogenic bacteria.

However, the good news is that the composition of the gut microbiome is a dynamic entity. With regular, sustained and long-term lifestyle modifications, the gut microbiome structure and function can be modified. We suggest you follow the recommendations highlighted in the book. Here are a few essential tips that can easily be incorporated to restore gut microbe health.

A. Include probiotics in your diet.

Probiotics have a restorative effect on gut microbiota. Instead of relying on just one source of probiotic, include diverse probiotics in the diet.

Here are a few examples- Yogurt, Kefir, Curd, Buttermilk, Sauerkraut, Tempeh, Kimchi, Miso, Kombucha, Pickles, and Natto.

B. Have fermented foods:

Fermented foods are excellent for gut health. Almost all the probiotics that we have discussed in the previous point fall under the category of fermented foods. Additionally, some of the traditional Indian fermented foods include Idli, Dosa, Appam, Dhokla, and Kanji, etc., which are known to impact gut microbiome structure and function in favor of good health.

C. Up your fiber intake:

As we have already learned upon reading the book, fiber in the diet is important for gut bacteria to thrive. A diet high in dietary fiber is associated with an increased diversity of the gut microbiome and favors the growth of beneficial microbial species. On the other hand, a diet devoid of fiber reduces gut diversity and fuels the growth of pathogenic bacteria and is a major contributor to the present-day lifestyle disease epidemic.

Here are a few strategies to incorporate fiber in the diet:

1. Have whole grains instead of refined grains. Whole grains are a rich source of fiber apart from being an excellent source of many vitamins and minerals such as Vitamin B12 and Iron.
2. Avoid consuming processed and packaged food. Such food is devoid of fiber and rich in preservatives, sodium, and refined sugar, all of which alter gut microbiome composition in an undesirable way.
3. Incorporate fruits in the diet as snacks. Ensure diversity in fruit selection.
4. Berries are a great source of dietary fiber. These are an excellent source of polyphenols, which help gut bacteria thrive.
5. Include peas and beans in your diet regularly.
6. Make nuts and seeds an integral part of your breakfast.
7. Have diverse raw and cooked vegetables regularly..

D. Consume prebiotics

In addition to taking probiotics, it is essential to incorporate prebiotics into the diet to maintain gut health. Simply described, prebiotics are what gut bacteria eat. For the probiotic bacteria to flourish, prebiotics are the fuel! Although the foods rich in fiber are generally a good source of prebiotics, there are foods that are not a particularly rich source of fiber but contain prebiotics in a good amount.

Some common examples of foods that are a rich source of prebiotics include garlic, onion, banana, barley, oats, apple, flaxseeds, wheat-bran, cocoa, asparagus, leeks, chicory Root, dandelion greens, and Jerusalem Artichoke.

This is an overview of the generalized dietary recommendations that help restore gut health after indiscriminate use of antibiotics, but apart from these, optimum sleep, stress management, exercise, physical activity, and an overall healthy lifestyle are critical to maintaining gut health.

Q: What is the difference between cardio and weight training exercises? Which type is good for gut-flora?

A- Based on the mode of the derivation of energy, exercise can either be aerobic or anaerobic. In aerobic exercises, the primary mode of getting the energy required for movement is through aerobic metabolism, which needs oxygen and utilizes predominantly fats and carbohydrates as the substrate. While the anaerobic exercises use carbohydrates stored as glycogen in the muscles to provide energy. Anaerobic exercises produce lactate. Common aerobic exercises such as brisk walk, running, cycling, swimming, rowing, etc. are popularly known as cardio. The weight-training exercises are anaerobic exercises and also known as strength training. As far as the healthy gut-flora is concerned, as already mentioned in chapter 1, the scientific evidence is primarily about the beneficial impact of cardio or aerobic exercises. But to derive optimal health benefits, a mix of both cardio and weight training exercises are recommended. You can choose one of these recommendations:

- At least 150 minutes of moderate aerobic activity such as cycling or brisk walk every week
  And
- Strength training exercises on two or more days a week that work all the major muscles (legs, hips, back, abdomen, chest, shoulders, and arms)
  OR
- 75 minutes of vigorous aerobic activity such as running every week
  And
- Strength training exercises on two or more days a week that work all the major muscles
  OR
- A mix of moderate and vigorous aerobic activity every week — for example, two 30-minute runs plus one 30 minutes brisk walking equates to 150 minutes of moderate aerobic activity
  And
- Strength training exercises on two or more days a week that work all the major muscles.

To summarize, although the scientific evidence to achieve a healthy gut-flora is mainly for cardio at present, one should do both the cardio and strength training for maximum benefits.

Q: I understand that physical activity is vital for gut health. I am a busy professional with hardly any time for it. How can I be physically active?

A- You're right in your understanding of the fact that physical activity is critical for maintaining gut health. It is important for us first to understand that physical activity and exercise are two related yet different terms, and both are important for achieving and maintaining gut health.

Physical activity, by definition, is any movement that our muscles carry out utilizing energy. It includes all the movements that we undertake on a daily basis, such as fetching a glass of water from the kitchen or cleaning the house without the intention of improving health. Exercise, in contrast with physical activity, is an intentional movement that is planned, structured, and repetitive. Exercise is usually undertaken with the end goal of improving health or maintaining physical fitness. There is a common misconception that as long as we exercise, we can do away without physical activity. To ensure optimum gut health, regular physical activity is as important as exercise, if not more. As per the latest research, for every 30 minutes of sitting, it is recommended to undertake movement for 3 minutes. Here are a few strategies that can be employed by those who are pressed for time that can be integrated into the everyday lifestyle for optimum health and well-being:

1. While riding a bus, consider walking till the second closest bus stop, instead of the nearest one.
2. Park your car as far as possible instead of looking for the spot closest to your home/office.
3. When using an elevator, use the stairs for the last or two floors.
4. Chose to stand instead of sitting at your desk. (Ask your HR about your company's standing desk policy). Maybe, you'll even inspire others to lead a healthy lifestyle.
5. Instead of the usual office, engage your clients/colleagues/co-workers in walk-and-talk meetings.
6. Walk when you talk-over-phone.
7. During the lunch break, if you finish early, take a post-lunch walk instead of a post-lunch nap.
8. If you travel by metro or any other form of public transport, offer your seat to someone who needs it more than you.

9. When with the kids, behave like kids. Walk/play with the children while talking about their day. It will help flourish both your gut bacteria as well as your relationships.
10. While watching the TV, undertake intermittent stretching exercises.
11. Instead of indulging in stress eating, bust your stress by a brisk walk, or better still, a casual jog.
12. If you need to travel less than 1 km, always walk, instead of using the car/scooter/motorbike. It takes just 10 minutes.
13. Instead of meeting friends over a coffee and snacks, invite them for a 'walk-in-the-park' session.

These small steps, followed daily, have the potential to bring about a significant difference to your everyday physical activity levels. These do not demand any time either. You may be 16 or 60 years old; it's not too late for you to start practicing these minor changes to experience better health.

Q: I was surprised to know that even sleep affects the composition and nature of our gut-flora? How many hours should I sleep? What are the tips for good quality sleep at night?

A- Yes, it is indeed surprising to know that something seemingly as unrelated as sleep can have an impact on bacteria living in our intestines. As far as hours of sleep are concerned, most scientific guidelines recommend having 7 hours of uninterrupted sleep every night for a healthy adult. Though there are long sleepers who may need up to 10 hours of night sleep to function optimally and short sleepers who are active and functional even with 6 hours of sleep.

The numbers of recommended hours of sleep also vary with age. A new-born baby sleeps as much as 16 to 20 hours a day. The sleep requirement gradually reduces with age. Senior citizens would not be able to sleep for more than 6 hours.

Good quality sleep means differently to different individuals. Nevertheless, one should feel rested, refreshed, relaxed, and should function optimally after a good sleep. Good sleep is an essential component of staying physically and mentally healthy. It is only during sleep that some of the most important hormones are released in the body, and the repair work is done. Even there is rejigging of memory from one part of the brain to another while we sleep, which is essential for the proper functioning of our memory.

One can follow SLEEP HYGIENE for good quality night's sleep. These practices are conducive to sleeping well on a regular basis. Sleep hygiene includes the following –

1. One must follow a time schedule for going to bed and waking up; especially the wake-up time needs to be fixed strictly.
2. One should avoid the daytime naps..
3. Exercise regularly, preferably in the first half of the day.
4. Avoid tea/coffee, tobacco, and other stimulants after late afternoon.
5. Have an early light dinner around 7 pm.
6. You can also take a hot shower before bed. It will be relaxing and help you in falling asleep.
7. Wear loose, comfortable cotton clothes in bed.
8. Avoid all types of screens in bed - phones, laptops, tablets, etc.
9. Do not engage in thrillers at bedtime - a light-hearted book is the way to go.
10. Do not spend more than 10 min in bed if you are not able to sleep. Do something like reading a book and get back only if you feel sleepy.
11. Bed should be used only for sleeping; avoid eating, reading, using laptops, and other such activities in bed.
12. Bedroom should have ambient light and temperature. Avoid a noisy room.

Q: You have described in the book how stress negatively impacts the gut-flora and how meditation helps restore it. Can you share some information about a few relaxation techniques that can be easily undertaken at home?

A- Absolutely! Taking up relaxation techniques significantly enhances overall well-being. Relaxation is considered as the "Aspirin" of Behavioral medicine. There are two reasons for this. The first is that it has many variants. The second is that it is relevant in treating many conditions (sleep disturbance, headache, hypertension, asthma, problematic alcohol usage, hyperactivity, and various forms of anxiety) apart from improving the general well-being.

Talking about the relaxation techniques, there are almost 26 empirical types. Some are core psychological techniques (cognitive-behavioral), while others are 'effective' techniques based on ancient traditions across the different parts of the world. These include Yoga, Tai Chi, and Lamaze, etc. Let us learn about a few popular cognitive-behavioral (psychological) techniques for relaxation. These include diaphragmatic breathing (breathing retraining), hypnosis, guided imagery, progressive muscle relaxation, biofeedback and neurofeedback, and mindfulness meditations, to name the most common ones.

- Diaphragmatic breathing involves breathing slow and deep, which promotes a subjective state of relaxation.
- Hypnosis is based on the principles of suggestibility, where an individual achieves more and more relaxation with the help of cued suggestions.
- Guided imagery exploits an individual's ability to visualize a cue associated with relaxation.
- Progressive muscle relaxation involves a muscular tense-release procedure targeting different muscle groups.
- Biofeedback and neurofeedback techniques are based on principles of "learning", wherein an individual gets real-time information/feedback about physiological and neurological

responses. The individual 'learns' to remain in an optimum state of arousal in response to the relaxation techniques parameters.

- Mindfulness utilizes the principles of "meditation," wherein an individual intentionally concentrates on any one aspect of the environment/body sensations with a non-judgmental attitude experiencing the present moment fully.

The standard traditional techniques are:

- Yoga is a way of living rather than just a relaxation technique. It has eight limbs, out of which *pranayama* and *asana* are particularly useful for relaxation. Pranayama includes a variety of breathing exercises during which a practitioner concentrates on his/her breathing. Asana basically means 'posture'. A large number of different postures are described ranging from simple ones to increasingly complex ones. A practitioner performs and holds the specific asana for a brief period of time, depending on his/her level of practice. These asana exercise all the muscles, nerves, and glands in the human body.
- Tai chi is the traditional internal Chinese martial art with dual applications- It is employed both as a means for self-defense as well as a relaxation technique. It is popularly described as meditation in motion, and now, due to its health benefits, it is also being called "*medication* in motion." Tai chi involves low-impact, slow-motion exercises that anyone can take up irrespective of health status and age.
- The Lamaze technique or Lamaze was popularized by the French obstetrician Dr. Fernand Lamaze during the early 1950s. It is currently one of the most popular birthing programs taken up by those inclined towards natural birth. The Lamaze technique aims to make birthing a comfortable experience and utilizes a combination of relaxation techniques, breathing exercises, movement, and massage.

There are many indications and contraindications of each technique. The choice also depends on one's personal preference and overall attitude towards a particular relaxation technique. Most of these techniques require assistance and training. It is not advisable to perform them without any guidance and supervision. We have just introduced to you a few of the most popular relaxation techniques. However, you must consult a practitioner, coach, or teacher to learn and practice them correctly.

Q. You have mentioned in the book that fermented foods work wonders in optimizing gut health. Can you elaborate more on the different types of fermented foods used by people across cultures globally?

A- It is quite interesting to know that fermented foods have been used by humans, across different cultures, for millennia. Our ancestors came to their independent conclusions that fermented food items are an essential determinant of gut health. Fermentation of food broadly refers to any process, which, with the activity of microorganisms, brings about a desirable change to food. It is vital to know about the different types of fermented food items used by cultures across the globe; however, due to the limited scope of this book, we will discuss the five popular ones consumed in different countries today.

1. Kanji is a popular beverage in the Indian subcontinent, used primarily as an appetizer during the winter months in India. It is made with water, black carrots, mustard seeds, and heeng (asafoetida). In the Indian subcontinent, black carrots grow during the winter months; the color of the purple and black varieties is due to a high concentration of anthocyanins, in contrast to a higher concentration of beta-carotene in "common" red carrots. Mustard seeds help in keeping the body warm in winters. When preparing, the drink is kept in the direct sunlight and allowed to undergo fermentation in glass or ceramic jars for 3-4 days.
2. Sauerkraut is a popular fermented dish, which, despite its German name, originated in central and eastern Europe. Sauerkraut, which

loosely translates to "sour cabbage", is made by process of pickling called lactic acid fermentation. Finely shredded cabbage is layered with salt and left to undergo fermentation. Fully cured sauerkraut can be kept for several months in an airtight container at 15°C (60°F) or below. Captain James Cook always kept a reserve of sauerkraut on his sea voyages, as he believed it prevented scurvy.

3. Kimchi is a traditional side dish in Korean cuisine. It consists of salted and fermented vegetables, such as Napa cabbage and Korean radish. It is made with a widely varying selection of seasonings such as *gochugaru* (chilli powder), ginger, garlic, spring onions, and *jeotgal* (seafood), etc. There are multiple variations of kimchi in different areas of the Korean peninsula, made with other vegetables as the main ingredients. Traditionally, it was stored in underground large earthen pots to prevent it from getting frozen during the winter months. Conversely, during the summer months, the kimchi stayed cool enough to slow down the fermentation because of the underground storage.
4. Miso is a traditional food item produced by fermenting soybeans with salt and using a unique fungus - kōji (*Aspergillus oryzae*) popular in Japan. It is a thick paste used as sauces and spreads. Sometimes rice, barley, seaweed, and other ingredients are also used. Miso soup called *misoshiru* is prepared by pickling vegetables or meats and mixing with the traditional soup broth in japan - dashi soup stock. It is a Japanese culinary staple, consumed daily by much of the Japanese population. The pairing of plain rice and miso soup is the fundamental unit of Japanese cuisine. Miso is widely used in Japan, both in traditional and modern cooking. The most common miso flavors are *Shiromiso*, *Akamiso*, and *Awasemiso,* respectively called white miso, red miso, and "mixed miso." Although white and red are the most common types available, different varieties are preferred in different geographic regions of Japan.

5. Tempeh or Tempe is a traditional Indonesian soy product, which is made from fermented soybeans. It is made by natural culturing and controlled fermentation that makes it into a cake form. A special fungus, *Rhizopus oligosporus,* locally called tempeh starter, is used for fermentation. Tempeh is made with whole soybeans, which are softened by soaking, the hull is removed, and it is partly cooked. It has a firm texture and an earthy flavor, which becomes more pronounced as it ages. Many tempeh variations exist, which may use other types of beans, wheat, or a mixture of both. The fermentation process and the use of the whole soybean give it a high protein, dietary fiber, and vitamin content. Tempeh is widely popular on the island of Java, where it is a staple source of protein.

All these fermented foods are great for optimizing gut health and also taste incredible. Worth a try for a cross-cultural eating experience!

Q: I am a 55-year-old woman. I have diabetes and hypertension, which are controlled on medicines prescribed by the physician. Can I switch over to probiotics?

A- No. Absolutely not! You must continue medicines under the care of your physician. Probiotics have shown some beneficial effects, especially in diabetes, but at present, the probiotics are not efficacious enough to be used as treatment. At best, probiotics can be used as an adjunct to the standard therapy. In no way can they replace the existing drugs for diabetes and hypertension, which are highly effective and safe at present. No guidelines recommend the use of probiotics for the treatment of these diseases. Even a Fecal Microbiota Transplant, which has a much more significant impact on gut-flora is not advisable for the cure of diabetes or hypertension at present.

BIBLIOGRAPHY

CHAPTER 1

Allen, J. M., Mailing, L. J., Niemiro, G. M., Moore, R., Cook, M. D., White, B. A., Holscher, H. D., & Woods, J. A. 2018. Exercise Alters Gut Microbiota Composition and Function in Lean and Obese Humans. *Medicine and science in sports and exercise*, 50(4): 747–57. https://doi.org/10.1249/MSS.0000000000001495.

Benedict, C., Vogel, H., Jonas, W., Woting, A., Blaut, M., Schürmann, A., &Cedernaes, J. 2016. Gut microbiota and glucometabolic alterations in response to recurrent partial sleep deprivation in normal-weight young individuals. *Molecular metabolism*, 5(12): 1175–86. https://doi.org/10.1016/j.molmet.2016.10.003.

Bezirtzoglou, E., Tsiotsias, A., & Welling, G. W. 2011. Microbiota profile in feces of breast- and formula-fed newborns by using fluorescence in situ hybridization (FISH). *Anaerobe*, 17(6): 478–82. https://doi.org/10.1016/j.anaerobe.2011.03.009.

Clarke, S. F., Murphy, E. F., O'Sullivan, O., Lucey, A. J., Humphreys, M., Hogan, A., Hayes, P., O'Reilly, M., Jeffery, I. B., Wood-Martin, R., Kerins, D. M., Quigley, E., Ross, R. P., O'Toole, P. W., Molloy, M. G., Falvey, E., Shanahan, F., & Cotter, P. D. 2014. Exercise and associated dietary extremes impact on gut microbial diversity. *Gut*, 63(12):1913–20. https://doi.org/10.1136/gutjnl-2013-306541.

Costantini, L., Molinari, R., Farinon, B., & Merendino, N. 2017. Impact of Omega-3 Fatty Acids on the Gut Microbiota. *International journal of molecular sciences,* 18(12): 2645. https://doi.org/10.3390/ijms18122645.

Davis, C., Bryan, J., Hodgson, J., & Murphy, K. 2015. Definition of the Mediterranean Diet; a Literature Review. *Nutrients*, 7(11):9139–53. https://doi.org/10.3390/nu7115459.

De Filippis, F., Pellegrini, N., Vannini, L., Jeffery, I. B., La Storia, A., Laghi, L., Serrazanetti, D. I., Di Cagno, R., Ferrocino, I., Lazzi, C., Turroni, S., Cocolin, L., Brigidi, P., Neviani, E., Gobbetti, M., O'Toole, P. W., &Ercolini, D. 2016. High-level adherence to a Mediterranean diet beneficially impacts the gut microbiota and associated metabolome. *Gut*, 65(11):1812–21. https://doi.org/10.1136/gutjnl-2015-309957.

Deaver, J. A., Eum, S. Y., &Toborek, M. 2018. Circadian Disruption Changes Gut Microbiome Taxa and Functional Gene Composition. *Frontiers in microbiology*, 9, 737. https://doi.org/10.3389/fmicb.2018.00737.

Dhakan, D. B., Maji, A., Sharma, A. K., Saxena, R., Pulikkan, J., Grace, T., Gomez, A., Scaria, J., Amato, K. R., & Sharma, V. K. 2019. The unique composition of Indian gut microbiome, gene catalogue, and associated fecalmetabolome deciphered using multi-omics approaches. *GigaScience*, 8(3): giz004. https://doi.org/10.1093/gigascience/giz004.

Dominguez-Bello, M. G., Costello, E. K., Contreras, M., Magris, M., Hidalgo, G., Fierer, N., & Knight, R. 2010. Delivery mode shapes the acquisition and structure of the initial microbiota across multiple body habitats in newborns. *Proceedings of the National Academy of Sciences of the United States of America*, 107(26): 11971–5. https://doi.org/10.1073/pnas.1002601107.

Estaki, M., Pither, J., Baumeister, P., Little, J. P., Gill, S. K., Ghosh, S., Ahmadi-Vand, Z., Marsden, K. R., & Gibson, D. L. 2016. Cardiorespiratory fitness as a predictor of intestinal microbial diversity and distinct metagenomic functions. *Microbiome*, 4(1), 42. https://| doi.org/10.1186/s40168-016-0189-7.

Galazzo, G., van Best, N., Bervoets, L., Dapaah, I. O., Savelkoul, P. H., Hornef, M. W., GI-MDH consortium, Lau, S., Hamelmann, E., &Penders, J. 2020. Development of the Microbiota and Associations With Birth Mode, Diet, and Atopic Disorders in a Longitudinal Analysis of Stool Samples, Collected From Infancy Through Early Childhood. *Gastroenterology*, 158(6): 1584–96. https://doi.org/10.1053/j.gastro.2020.01.024.

Imhann, F., Bonder, M. J., Vich Vila, A., Fu, J., Mujagic, Z., Vork, L., Tigchelaar, E. F., Jankipersadsing, S. A., Cenit, M. C., Harmsen, H. J., Dijkstra, G., Franke, L., Xavier, R. J., Jonkers, D., Wijmenga, C., Weersma, R. K., & Zhernakova, A. 2016. Proton pump inhibitors affect the gut microbiome. *Gut*, 65(5): 740–8. https://doi.org/10.1136/gutjnl-2015-310376.

Jung, J. Y., Lee, S. H., Kim, J. M., Park, M. S., Bae, J. W., Hahn, Y., Madsen, E. L., &Jeon, C. O. 2011. Metagenomic analysis of kimchi, a traditional Korean fermented food. *Applied and environmental microbiology*, 77(7): 2264–74. https://doi.org/10.1128/AEM.02157-10.

Kromhout, D., Menotti, A., Bloemberg, B., Aravanis, C., Blackburn, H., Buzina, R., Dontas, A. S., Fidanza, F., Giampaoli, S., & Jansen, A. 1995. Dietary saturated and trans fatty acids and cholesterol and 25-year mortality from coronary heart disease: the Seven Countries Study. *Preventive medicine*, 24(3): 308–15. https://doi.org/10.1006/pmed.1995.1049.

Laursen, M. F., Zachariassen, G., Bahl, M. I., Bergström, A., Høst, A., Michaelsen, K. F., &Licht, T. R. 2015. Having older siblings is associated with gut microbiota development during early childhood. *BMC microbiology*, 15: 154. https://doi.org/10.1186/s12866-015-0477-6.

Ma, G & Chen, Y. 2020. Polyphenol supplementation benefits human health via gut microbiota: A systematic review via meta-analysis. *Journal of Functional Foods*. 66. 103829. 10.1016/j.jff.2020.103829.

Madsen, L., Myrmel, L. S., Fjære, E., Liaset, B., & Kristiansen, K. 2017. Links between Dietary Protein Sources, the Gut Microbiota, and Obesity. *Frontiers in physiology*, 8: 1047. https://doi.org/10.3389/fphys.2017.01047.

Moreno-Pérez, D., Bressa, C., Bailén, M., Hamed-Bousdar, S., Naclerio, F., Carmona, M., Pérez, M., González-Soltero, R., Montalvo-Lominchar, M. G., Carabaña, C., & Larrosa, M. (2018). Effect of a Protein Supplement on the Gut Microbiota of Endurance Athletes: A Randomized, Controlled, Double-Blind Pilot Study. *Nutrients*, 10(3), 337. https://doi.org/10.3390/nu10030337.

Madsen, L., Myrmel, L. S., Fjære, E., Liaset, B., & Kristiansen, K. 2017. Links between Dietary Protein Sources, the Gut Microbiota, and Obesity. *Frontiers in physiology*, 8: 1047. https://doi.org/10.3389/fphys.2017.01047.

Maier, L., Pruteanu, M., Kuhn, M., Zeller, G., Telzerow, A., Anderson, E. E., Brochado, A. R., Fernandez, K. C., Dose, H., Mori, H., Patil, K. R., Bork, P., & Typas, A. 2018. Extensive impact of non-antibiotic drugs on human gut bacteria. *Nature*, 555(7698): 623–8. https://doi.org/10.1038/nature25979.

Matsumoto, M., Inoue, R., Tsukahara, T., Ushida, K., Chiji, H., Matsubara, N., & Hara, H. 2008. Voluntary running exercise alters microbiota composition and increases n-butyrate concentration in the rat cecum. *Bioscience, biotechnology, and biochemistry*, 72(2): 572–6. https://doi.org/10.1271/bbb.70474.

Mitsou, E. K., Kakali, A., Antonopoulou, S., Mountzouris, K. C., Yannakoulia, M., Panagiotakos, D. B., & Kyriacou, A. 2017. Adherence to the Mediterranean diet is associated with the gut microbiota pattern and gastrointestinal characteristics in an adult population. *The British journal of nutrition*, 117(12): 1645–55. https://doi.org/10.1017/S0007114517001593.

Moszak, M., Szulińska, M., &Bogdański, P. 2020. You Are What You Eat-The Relationship between Diet, Microbiota, and Metabolic Disorders-A Review. *Nutrients*, 12(4): 1096. https://doi.org/10.3390/nu12041096.

Rogers, M., & Aronoff, D. M. 2016. The influence of non-steroidal anti-inflammatory drugs on the gut microbiome. *Clinical microbiology and infection: the official publication of the European Society of Clinical Microbiology and Infectious Diseases*, 22(2): 178.e1–178.e9. https://doi.org/10.1016/j.cmi.2015.10.003.

Rutayisire, E., Huang, K., Liu, Y., & Tao, F. 2016. The mode of delivery affects the diversity and colonization pattern of the gut microbiota during the first year of infants' life: a systematic review. *BMC gastroenterology*, 16(1): 86. https://doi.org/10.1186/s12876-016-0498-0.

Shin, J. H., Jung, S., Kim, S. A., Kang, M. S., Kim, M. S., Joung, H., Hwang, G. S., & Shin, D. M. 2019. Differential Effects of Typical Korean Versus American-Style Diets on Gut Microbial Composition and Metabolic Profile in Healthy Overweight Koreans: A Randomized Crossover Trial. *Nutrients*, 11(10): 2450. https://doi.org/10.3390/nu11102450.

Smits, S. A., Leach, J., Sonnenburg, E. D., Gonzalez, C. G., Lichtman, J. S., Reid, G., Knight, R., Manjurano, A., Changalucha, J., Elias, J. E., Dominguez-Bello, M. G., & Sonnenburg, J. L. 2017. Seasonal cycling in the gut microbiome of the Hadza hunter-gatherers of Tanzania. *Science (New York, N.Y.)*, 357(6353): 802–6. https://doi.org/10.1126/science.aan4834.

Subbarao, P., Anand, S. S., Becker, A. B., Befus, A. D., Brauer, M., Brook, J. R., Denburg, J. A., HayGlass, K. T., Kobor, M. S., Kollmann, T. R., Kozyrskyj, A. L., Lou, W. Y., Mandhane, P. J., Miller, G. E., Moraes, T. J., Pare, P. D., Scott, J. A., Takaro, T. K., Turvey, S. E., Duncan, J. M., … CHILD Study investigators 2015. The Canadian Healthy Infant Longitudinal Development (CHILD) Study: examining developmental origins of allergy and asthma. *Thorax*, 70(10): 998–1000. https://doi.org/10.1136/thoraxjnl-2015-207246.

Thorburn, A. N., McKenzie, C. I., Shen, S., Stanley, D., Macia, L., Mason, L. J., Roberts, L. K., Wong, C. H., Shim, R., Robert, R., Chevalier, N., Tan, J. K., Mariño, E., Moore, R. J., Wong, L., McConville, M. J., Tull, D. L., Wood, L. G., Murphy, V. E., Mattes, J., Mackay, C. R.

2015. Evidence that asthma is a developmental origin disease influenced by maternal diet and bacterial metabolites. *Nature communications*, 6: 7320. https://doi.org/10.1038/ncomms8320.

Wu, H., Esteve, E., Tremaroli, V., Khan, M. T., Caesar, R., Mannerås-Holm, L., Ståhlman, M., Olsson, L. M., Serino, M., Planas-Fèlix, M., Xifra, G., Mercader, J. M., Torrents, D., Burcelin, R., Ricart, W., Perkins, R., Fernàndez-Real, J. M., & Bäckhed, F. 2017. Metformin alters the gut microbiome of individuals with treatment-naive type 2 diabetes, contributing to the therapeutic effects of the drug. *Nature medicine*, 23(7): 850–8. https://doi.org/10.1038/nm.4345.

Xu, C., Zhu, H., &Qiu, P. 2019. Aging progression of human gut microbiota. *BMC microbiology*, 19(1): 236. https://doi.org/10.1186/s12866-019-1616-2.

Younge, N., McCann, J. R., Ballard, J., Plunkett, C., Akhtar, S., Araújo-Pérez, F., Murtha, A., Brandon, D., & Seed, P. C. 2019. Fetal exposure to the maternal microbiota in humans and mice. *JCI insight*, 4(19): e127806. https://doi.org/10.1172/jci.insight.127806.

CHAPTER 2

Bäckhed, F., Ding, H., Wang, T., Hooper, L. V., Koh, G. Y., Nagy, A., Semenkovich, C. F., & Gordon, J. I. 2004. The gut microbiota as an environmental factor that regulates fat storage. *Proceedings of the National Academy of Sciences of the United States of America*, 101(44): 15718–23. https://doi.org/10.1073/pnas.0407076101.

Dzidic, M., Boix-Amorós, A., Selma-Royo, M., Mira, A., &Collado, M. C. 2018. Gut Microbiota and Mucosal Immunity in the Neonate. *Medical sciences (Basel, Switzerland)*, 6(3): 56. https://doi.org/10.3390/medsci6030056.

Galazzo, G., van Best, N., Bervoets, L., Dapaah, I. O., Savelkoul, P. H., Hornef, M. W., GI-MDH consortium, Lau, S., Hamelmann, E., & Penders, J. 2020. Development of the Microbiota and Associations With Birth Mode, Diet, and Atopic Disorders in a Longitudinal

Analysis of Stool Samples, Collected From Infancy Through Early Childhood. *Gastroenterology*, 158(6): 1584–96. https://doi.org/10.1053/j.gastro.2020.01.024.

Hamer, H. M., Jonkers, D., Venema, K., Vanhoutvin, S., Troost, F. J., &Brummer, R. J. 2008. Review article: the role of butyrate on colonic function. *Alimentary pharmacology & therapeutics*, 27(2): 104–19. https://doi.org/10.1111/j.1365-2036.2007.03562.x.

Konieczna, P., Groeger, D., Ziegler, M., Frei, R., Ferstl, R., Shanahan, F., Quigley, E. M., Kiely, B., Akdis, C. A., & O'Mahony, L. 2012. Bifidobacterium infantis 35624 administration induces Foxp3 T regulatory cells in human peripheral blood: potential role for myeloid and plasmacytoid dendritic cells. *Gut*, 61(3): 354–66. https://doi.org/10.1136/gutjnl-2011-300936.

Li, Z., Yi, C. X., Katiraei, S., Kooijman, S., Zhou, E., Chung, C. K., Gao, Y., van den Heuvel, J. K., Meijer, O. C., Berbée, J., Heijink, M., Giera, M., Willems van Dijk, K., Groen, A. K., Rensen, P., & Wang, Y. 2018. Butyrate reduces appetite and activates brown adipose tissue via the gut-brain neural circuit. *Gut*, 67(7): 1269–79. https://doi.org/10.1136/gutjnl-2017-314050.

Luby, S. P., Agboatwalla, M., Feikin, D. R., Painter, J., Billhimer, W., Altaf, A., & Hoekstra, R. M. 2005. Effect of handwashing on child health: a randomised controlled trial. *Lancet (London, England)*, 366(9481): 225–33. https://doi.org/10.1016/S0140-6736(05)66912-7.

Martens, E. C., Chiang, H. C., & Gordon, J. I. 2008. Mucosal glycan foraging enhances fitness and transmission of a saccharolytic human gut bacterial symbiont. *Cell host & microbe*, 4(5): 447–57. https://doi.org/10.1016/j.chom.2008.09.007.

Peng, L., Li, Z. R., Green, R. S., Holzman, I. R., & Lin, J. 2009. Butyrate enhances the intestinal barrier by facilitating tight junction assembly via activation of AMP-activated protein kinase in Caco-2 cell monolayers. *The Journal of nutrition*, 139(9): 1619–25. https://doi.org/10.3945/jn.109.104638.

Shimotoyodome, A., Meguro, S., Hase, T., Tokimitsu, I., & Sakata, T. 2000. Short chain fatty acids but not lactate or succinate stimulate mucus release in the rat colon. *Comparative biochemistry and physiology. Part A, Molecular & integrative physiology*, 125(4): 525–31. https://doi.org/10.1016/s1095-6433(00)00183-5.

Smits, H. H., Engering, A., van der Kleij, D., de Jong, E. C., Schipper, K., van Capel, T. M., Zaat, B. A., Yazdanbakhsh, M., Wierenga, E. A., van Kooyk, Y., &Kapsenberg, M. L. 2005. Selective probiotic bacteria induce IL-10-producing regulatory T cells in vitro by modulating dendritic cell function through dendritic cell-specific intercellular adhesion molecule 3-grabbing nonintegrin. *The Journal of allergy and clinical immunology*, 115(6):1260–67. https://doi.org/10.1016/j.jaci.2005.03.036.

Thomas, JP., Parker, A., Divekar, D., Carmen, P., Watson, A,. 2018. PTU-066 The gut microbiota influences intestinal epithelial proliferative potential. *Gut*, 67:A204.

Zeng, H., & Chi, H. 2015. Metabolic control of regulatory T cell development and function. *Trends in immunology*, 36(1): 3–12. https://doi.org/10.1016/j.it.2014.08.003.

CHAPTER 3

Alang, N., & Kelly, C. R. 2015. Weight gain after fecalmicrobiota transplantation. *Open forum infectious diseases*, *2*(1), ofv004. https://doi.org/10.1093/ofid/ofv004.

Balamurugan, R., George, G., Kabeerdoss, J., Hepsiba, J., Chandragunasekaran, A. M., & Ramakrishna, B. S. 2010. Quantitative differences in intestinal Faecalibacteriumprausnitzii in obese Indian children. *The British journal of nutrition*, 103(3): 335–38. https://doi.org/10.1017/S0007114509992182.

Cani, P., Amar, J., Iglesias, M., Poggi, M., Knauf, C., Bastelica, D., Neyrinck, A., Fava, F., Tuohy, K., Chabo, C., Waget, A., Delmee, E., Cousin, B., Sulpice, T., Chamontin, B., Ferrieres, J., Tanti, J., Gibson, G., Casteilla, L., Delzenne, N., Alessi, M. and Burcelin, R., 2007.

Metabolic Endotoxemia Initiates Obesity and Insulin Resistance. *Diabetes*, 56(7):1761-72.

Craven, L., Rahman, A., Nair Parvathy, S., Beaton, M., Silverman, J., Qumosani, K., Hramiak, I., Hegele, R., Joy, T., Meddings, J., Urquhart, B., Harvie, R., McKenzie, C., Summers, K., Reid, G., Burton, J. P., & Silverman, M. 2020. Allogenic Fecal Microbiota Transplantation in Patients With Nonalcoholic Fatty Liver Disease Improves Abnormal Small Intestinal Permeability: A Randomized Control Trial. *The American journal of gastroenterology*, 115(7):1055–65. https://doi.org/10.14309/ajg.0000000000000661.

Kalliomäki, M., Collado, M. C., Salminen, S., &Isolauri, E. 2008. Early differences in fecalmicrobiota composition in children may predict overweight. *The American journal of clinical nutrition*, 87(3): 534–38. https://doi.org/10.1093/ajcn/87.3.534.

Kootte, R. S., Levin, E., Salojärvi, J., Smits, L. P., Hartstra, A. V., Udayappan, S. D., Hermes, G., Bouter, K. E., Koopen, A. M., Holst, J. J., Knop, F. K., Blaak, E. E., Zhao, J., Smidt, H., Harms, A. C., Hankemeijer, T., Bergman, J., Romijn, H. A., Schaap, F. G., OldeDamink, S., … Nieuwdorp, M. 2017. Improvement of Insulin Sensitivity after Lean Donor Feces in Metabolic Syndrome Is Driven by Baseline Intestinal Microbiota Composition. *Cell metabolism*, 26(4): 611–19.e6. https://doi.org/10.1016/j.cmet.2017.09.008.

Larsen, N., Vogensen, F. K., van den Berg, F. W., Nielsen, D. S., Andreasen, A. S., Pedersen, B. K., Al-Soud, W. A., Sørensen, S. J., Hansen, L. H., & Jakobsen, M. 2010. Gut microbiota in human adults with type 2 diabetes differs from non-diabetic adults. *PloS one*, 5(2): e9085. https://doi.org/10.1371/journal.pone.0009085.

Qin, J., Li, Y., Cai, Z., Li, S., Zhu, J., Zhang, F., Liang, S., Zhang, W., Guan, Y., Shen, D., Peng, Y., Zhang, D., Jie, Z., Wu, W., Qin, Y., Xue, W., Li, J., Han, L., Lu, D., Wu, P., … Wang, J. 2012. A metagenome-wide association study of gut microbiota in type 2 diabetes. *Nature*, 490 (7418): 55–60. https://doi.org/10.1038/nature11450.

Rabot, S., Membrez, M., Bruneau, A., Gérard, P., Harach, T., Moser, M., Raymond, F., Mansourian, R., & Chou, C. J. 2010. Germ-free C57BL/6J mice are resistant to high-fat-diet-induced insulin resistance and have altered cholesterol metabolism. *FASEB journal: official publication of the Federation of American Societies for Experimental Biology*, 24(12): 4948–59. https://doi.org/10.1096/fj.10-164921.

Tao, Y. W., Gu, Y. L., Mao, X. Q., Zhang, L., & Pei, Y. F. 2020. Effects of probiotics on type II diabetes mellitus: a meta-analysis. *Journal of translational medicine*, 18(1): 30. https://doi.org/10.1186/s12967-020-02213-2.

Turnbaugh, P. J., Ley, R. E., Mahowald, M. A., Magrini, V., Mardis, E. R., & Gordon, J. I. 2006. An obesity-associated gut microbiome with increased capacity for energy harvest. *Nature*, 444(7122): 1027–31. https://doi.org/10.1038/nature05414.

Vrieze, A., Van Nood, E., Holleman, F., Salojärvi, J., Kootte, R. S., Bartelsman, J. F., Dallinga-Thie, G. M., Ackermans, M. T., Serlie, M. J., Oozeer, R., Derrien, M., Druesne, A., Van Hylckama Vlieg, J. E., Bloks, V. W., Groen, A. K., Heilig, H. G., Zoetendal, E. G., Stroes, E. S., de Vos, W. M., Hoekstra, J. B., … Nieuwdorp, M. 2012. Transfer of intestinal microbiota from lean donors increases insulin sensitivity in individuals with metabolic syndrome. *Gastroenterology*, 143(4): 913–6.e7. https://doi.org/10.1053/j.gastro.2012.06.031.

Wu, H., Tremaroli, V., Schmidt, C., Lundqvist, A., Olsson, L. M., Krämer, M., Gummesson, A., Perkins, R., Bergström, G., & Bäckhed, F. 2020. The Gut Microbiota in Prediabetes and Diabetes: A Population-Based Cross-Sectional Study. *Cell metabolism*, 32(3): 379–90.e3. https://doi.org/10.1016/j.cmet.2020.06.01.

Yu, E. W., Gao, L., Stastka, P., Cheney, M. C., Mahabamunuge, J., Torres Soto, M., Ford, C. B., Bryant, J. A., Henn, M. R., & Hohmann, E. L. 2020. Fecal microbiota transplantation for the improvement of metabolism in obesity: The FMT-TRIM double-blind placebo-controlled pilot trial. *PLoS medicine*, 17(3): e1003051. https://doi.org/10.1371/journal.pmed.1003051.

CHAPTER 4

Adams, J. B., Johansen, L. J., Powell, L. D., Quig, D., & Rubin, R. A. 2011. Gastrointestinal flora and gastrointestinal status in children with autism--comparisons to typical children and correlation with autism severity. *BMC gastroenterology*, 11: 22. https://doi.org/10.1186/1471-230X-11-22.

Bercik, P., Denou, E., Collins, J., Jackson, W., Lu, J., Jury, J., Deng, Y., Blennerhassett, P., Macri, J., McCoy, K. D., Verdu, E. F., & Collins, S. M. 2011. The intestinal microbiota affect central levels of brain-derived neurotropic factor and behavior in mice. *Gastroenterology*, 141(2): 599–609.e6093. https://doi.org/10.1053/j.gastro.2011.04.052.

Bridgewater, L. C., Zhang, C., Wu, Y., Hu, W., Zhang, Q., Wang, J., Li, S., & Zhao, L. 2017. Gender-based differences in host behavior and gut microbiota composition in response to high fat diet and stress in a mouse model. *Scientific reports*, 7(1): 10776. https://doi.org/10.1038/s41598-017-11069-4.

Callaghan, B. L., Fields, A., Gee, D. G., Gabard-Durnam, L., Caldera, C., Humphreys, K. L., Goff, B., Flannery, J., Telzer, E. H., Shapiro, M., & Tottenham, N. 2020. Mind and gut: Associations between mood and gastrointestinal distress in children exposed to adversity. *Development and psychopathology*, 32(1): 309–28. https://doi.org/10.1017/S0954579419000087.

Chu, C., Murdock, M. H., Jing, D., Won, T. H., Chung, H., Kressel, A. M., Tsaava, T., Addorisio, M. E., Putzel, G. G., Zhou, L., Bessman, N. J., Yang, R., Moriyama, S., Parkhurst, C. N., Li, A., Meyer, H. C., Teng, F., Chavan, S. S., Tracey, K. J., Regev, A., … Artis, D. 2019. The microbiota regulate neuronal function and fear extinction learning. *Nature*, 574(7779): 543–48. https://doi.org/10.1038/s41586-019-1644-y.

Claesson, M. J., Jeffery, I. B., Conde, S., Power, S. E., O'Connor, E. M., Cusack, S., Harris, H. M., Coakley, M., Lakshminarayanan, B., O'Sullivan, O., Fitzgerald, G. F., Deane, J., O'Connor, M., Harnedy, N., O'Connor, K., O'Mahony, D., van Sinderen, D., Wallace, M.,

Brennan, L., Stanton, C., ... O'Toole, P. W. 2012. Gut microbiota composition correlates with diet and health in the elderly. *Nature*, 488(7410): 178–84. https://doi.org/10.1038/nature11319.

Desbonnet, L., Clarke, G., Shanahan, F., Dinan, T. G., & Cryan, J. F. 2014. Microbiota is essential for social development in the mouse. *Molecular psychiatry*, 19(2): 146–8. https://doi.org/10.1038/mp.2013.65.

Desbonnet, L., Garrett, L., Clarke, G., Bienenstock, J., & Dinan, T. G. 2008. The probiotic Bifidobacteria infantis: An assessment of potential antidepressant properties in the rat. *Journal of psychiatric research*, 43(2): 164–74. https://doi.org/10.1016/j.jpsychires.2008.03.009.

Dupont HL. Review article: evidence for the role of gut microbiota in irritable bowel syndrome and its potential influence on therapeutic targets. *Aliment Pharmacol Ther.* 2014 May; 39(10):1033-42. doi: 10.1111/apt.12728. Epub 2014 Mar 25. PMID: 24665829.

Gareau, M. G., Wine, E., Rodrigues, D. M., Cho, J. H., Whary, M. T., Philpott, D. J., Macqueen, G., & Sherman, P. M. 2011. Bacterial infection causes stress-induced memory dysfunction in mice. *Gut*, 60(3): 307–17. https://doi.org/10.1136/gut.2009.202515.

Jiang, H., Ling, Z., Zhang, Y., Mao, H., Ma, Z., Yin, Y., Wang, W., Tang, W., Tan, Z., Shi, J., Li, L., & Ruan, B. 2015. Altered fecal microbiota composition in patients with major depressive disorder. *Brain, behavior, and immunity*, 48: 186–94. https://doi.org/10.1016/j.bbi.2015.03.016.

Johnson, K. V. A. 2020. Gut microbiome composition and diversity are related to human personality traits. *Human Microbiome Journal,* 15:100069 https://doi.org/10.1016/j.humic.2019.100069.

Kang, D. W., Adams, J. B., Gregory, A. C., Borody, T., Chittick, L., Fasano, A., Khoruts, A., Geis, E., Maldonado, J., McDonough-Means, S., Pollard, E. L., Roux, S., Sadowsky, M. J., Lipson, K. S., Sullivan, M. B., Caporaso, J. G., & Krajmalnik-Brown, R. 2017. Microbiota Transfer Therapy alters gut ecosystem and improves gastrointestinal and autism symptoms: an open-label study. *Microbiome*, 5(1): 10. https://doi.org/10.1186/s40168-016-0225-7.

Kim, H. N., Yun, Y., Ryu, S., Chang, Y., Kwon, M. J., Cho, J., Shin, H., & Kim, H. L. 2018. Correlation between gut microbiota and personality in adults: A cross-sectional study. *Brain, behavior, and immunity*, 69: 374–85. https://doi.org/10.1016/j.bbi.2017.12.012.

Ng, Q. X., Loke, W., Venkatanarayanan, N., Lim, D. Y., Soh, A., & Yeo, W. S. 2019. A Systematic Review of the Role of Prebiotics and Probiotics in Autism Spectrum Disorders. *Medicina (Kaunas, Lithuania)*, 55(5): 129. https://doi.org/10.3390/medicina55050129.

O'Mahony, S. M., Marchesi, J. R., Scully, P., Codling, C., Ceolho, A. M., Quigley, E. M., Cryan, J. F., & Dinan, T. G. 2009. Early life stress alters behavior, immunity, and microbiota in rats: implications for irritable bowel syndrome and psychiatric illnesses. *Biological psychiatry*, 65(3): 263–67. https://doi.org/10.1016/j.biopsych.2008.06.026.

Sandler, R. H., Finegold, S. M., Bolte, E. R., Buchanan, C. P., Maxwell, A. P., Väisänen, M. L., Nelson, M. N., & Wexler, H. M. 2000. Short-term benefit from oral vancomycin treatment of regressive-onset autism. *Journal of child neurology*, 15(7): 429–35. https://doi.org/10.1177/088307380001500701.

Sharon, G., Cruz, N. J., Kang, D. W., Gandal, M. J., Wang, B., Kim, Y. M., Zink, E. M., Casey, C. P., Taylor, B. C., Lane, C. J., Bramer, L. M., Isern, N. G., Hoyt, D. W., Noecker, C., Sweredoski, M. J., Moradian, A., Borenstein, E., Jansson, J. K., Knight, R., Metz, T. O., … Mazmanian, S. K. 2019. Human Gut Microbiota from Autism Spectrum Disorder Promote Behavioral Symptoms in Mice. *Cell*, 177(6): 1600–18.e17. https://doi.org/10.1016/j.cell.2019.05.004.

Sudo, N., Chida, Y., Aiba, Y., Sonoda, J., Oyama, N., Yu, X. N., Kubo, C., & Koga, Y. 2004. Postnatal microbial colonization programs the hypothalamic-pituitary-adrenal system for stress response in mice. *The Journal of physiology*, 558(Pt 1): 263–75. https://doi.org/10.1113/jphysiol.2004.063388.

Szczesniak, O., Hestad, K. A., Hanssen, J. F., & Rudi, K. 2016. Isovaleric acid in stool correlates with human depression. *Nutritional neuroscience*, 19(7): 279–83. https://doi.org/10.1179/1476830515Y.0000000007.

Valles-Colomer, M., Falony, G., Darzi, Y., Tigchelaar, E. F., Wang, J., Tito, R. Y., Schiweck, C., Kurilshikov, A., Joossens, M., Wijmenga, C., Claes, S., Van Oudenhove, L., Zhernakova, A., Vieira-Silva, S., & Raes, J. 2019. The neuroactive potential of the human gut microbiota in quality of life and depression. *Nature microbiology*, 4(4): 623–32. https://doi.org/10.1038/s41564-018-0337-x.

van de Wouw, M., Boehme, M., Lyte, J. M., Wiley, N., Strain, C., O'Sullivan, O., Clarke, G., Stanton, C., Dinan, T. G., & Cryan, J. F. 2018. Short-chain fatty acids: microbial metabolites that alleviate stress-induced brain-gut axis alterations. *The Journal of physiology*, 596(20): 4923–44. https://doi.org/10.1113/JP276431.

Wallace, C., & Milev, R. 2017. The effects of probiotics on depressive symptoms in humans: a systematic review. *Annals of general psychiatry*, 16: 14. https://doi.org/10.1186/s12991-017-0138-2.

Williams, E. A., Stimpson, J., Wang, D., Plummer, S., Garaiova, I., Barker, M. E., & Corfe, B. M. 2009. Clinical trial: a multistrain probiotic preparation significantly reduces symptoms of irritable bowel syndrome in a double-blind placebo-controlled study. *Alimentary pharmacology & therapeutics*, 29(1): 97–103. https://doi.org/10.1111/j.1365-2036.2008.03848.x.

Yano, J. M., Yu, K., Donaldson, G. P., Shastri, G. G., Ann, P., Ma, L., Nagler, C. R., Ismagilov, R. F., Mazmanian, S. K., & Hsiao, E. Y. 2015. Indigenous bacteria from the gut microbiota regulate host serotonin biosynthesis. *Cell*, 161(2): 264–76. https://doi.org/10.1016/j.cell.2015.02.047.

Zijlmans, M. A., Korpela, K., Riksen-Walraven, J. M., de Vos, W. M., & de Weerth, C. 2015. Maternal prenatal stress is associated with the infant intestinal microbiota. *Psychoneuroendocrinology*, 53: 233–45. https://doi.org/10.1016/j.psyneuen.2015.01.006.

CHAPTER 5

Balamurugan, R., George, G., Kabeerdoss, J., Hepsiba, J., Chandragunasekaran, A. M., &Ramakrishna, B. S. 2010. Quantitative differences in intestinal Faecalibacterium prausnitzii in obese Indian children. *The British journal of nutrition*, 103(3): 335–38. https://doi.org/10.1017/S0007114509992182.

Bezirtzoglou, E., Tsiotsias, A., & Welling, G. W. 2011. Microbiota profile in feces of breast- and formula-fed newborns by using fluorescence in situ hybridization (FISH). *Anaerobe*, 17(6): 478–82. https://doi.org/10.1016/j.anaerobe.2011.03.009.

Boutagy, N. E., McMillan, R. P., Frisard, M. I., & Hulver, M. W. 2016. Metabolic endotoxemia with obesity: Is it real and is it relevant?. *Biochimie*, 124: 11–20. https://doi.org/10.1016/j.biochi.2015.06.020.

Kalliomäki, M., Collado, M. C., Salminen, S., &Isolauri, E. 2008. Early differences in fecalmicrobiota composition in children may predict overweight. *The American journal of clinical nutrition*, 87(3): 534–38. https://doi.org/10.1093/ajcn/87.3.534.

Larsen, N., Vogensen, F. K., van den Berg, F. W., Nielsen, D. S., Andreasen, A. S., Pedersen, B. K., Al-Soud, W. A., Sørensen, S. J., Hansen, L. H., &Jakobsen, M. 2010. Gut microbiota in human adults with type 2 diabetes differs from non-diabetic adults. *PloS one*, 5(2): e9085. https://doi.org/10.1371/journal.pone.0009085.

Magne, F., Gotteland, M., Gauthier, L., Zazueta, A., Pesoa, S., Navarrete, P., & Balamurugan, R. 2020. The Firmicutes/Bacteroidetes Ratio: A Relevant Marker of Gut Dysbiosis in Obese Patients?. *Nutrients*, 12(5): 1474. https://doi.org/10.3390/nu12051474.

Nishijima, S., Suda, W., Oshima, K., Kim, S. W., Hirose, Y., Morita, H., & Hattori, M. 2016. The gut microbiome of healthy Japanese and its microbial and functional uniqueness. *DNA research: an international journal for rapid publication of reports on genes and genomes*, 23(2), 125–33. https://doi.org/10.1093/dnares/dsw002.

CHAPTER 6

Biedermann, L., Zeitz, J., Mwinyi, J., Sutter-Minder, E., Rehman, A., Ott, S. J., Steurer-Stey, C., Frei, A., Frei, P., Scharl, M., Loessner, M. J., Vavricka, S. R., Fried, M., Schreiber, S., Schuppler, M., & Rogler, G. 2013. Smoking cessation induces profound changes in the composition of the intestinal microbiota in humans. *PloS one*, 8(3): e59260. https://doi.org/10.1371/journal.pone.0059260.

Bibbò, S., Settanni, C. R., Porcari, S., Bocchino, E., Ianiro, G., Cammarota, G., & Gasbarrini, A. 2020. Fecal Microbiota Transplantation: Screening and Selection to Choose the Optimal Donor. *Journal of clinical medicine*, 9(6): 1757. https://doi.org/10.3390/jcm9061757.

Choi, I. H., Noh, J. S., Han, J. S., Kim, H. J., Han, E. S., & Song, Y. O. 2013. Kimchi, a fermented vegetable, improves serum lipid profiles in healthy young adults: randomized clinical trial. *Journal of medicinal food*, 16(3): 223–9. https://doi.org/10.1089/jmf.2012.2563.

de Wit, N., Derrien, M., Bosch-Vermeulen, H., Oosterink, E., Keshtkar, S., Duval, C., de Vogel-van den Bosch, J., Kleerebezem, M., Müller, M., & van der Meer, R. 2012. Saturated fat stimulates obesity and hepatic steatosis and affects gut microbiota composition by an enhanced overflow of dietary fat to the distal intestine. *American journal of physiology. Gastrointestinal and liver physiology*, 303(5): G589–G99. https://doi.org/10.1152/ajpgi.00488.2011.

Duplessis, C. A., You, D., Johnson, M., & Speziale, A. 2012. Efficacious outcome employing fecal bacteriotherapy in severe Crohn's colitis complicated by refractory Clostridium difficile infection. *Infection*, 40(4):469–72. https://doi.org/10.1007/s15010-011-0226-1.

Ford, A. C., Quigley, E. M., Lacy, B. E., Lembo, A. J., Saito, Y. A., Schiller, L. R., Soffer, E. E., Spiegel, B. M., & Moayyedi, P. 2014. Efficacy of prebiotics, probiotics, and synbiotics in irritable bowel syndrome and chronic idiopathic constipation: systematic review and meta-analysis. *The American journal of gastroenterology*, 109(10): 1547–62. https://doi.org/10.1038/ajg.2014.202.

Goldenberg, J. Z., Lytvyn, L., Steurich, J., Parkin, P., Mahant, S., & Johnston, B. C. 2015. Probiotics for the prevention of pediatric antibiotic-associated diarrhea. *The Cochrane database of systematic reviews*, (12): CD004827. https://doi.org/10.1002/14651858.CD004827.pub4.

He, C., Wu, Q., Hayashi, N., Nakano, F., Nakatsukasa, E., & Tsuduki, T. 2020. Carbohydrate-restricted diet alters the gut microbiota, promotes senescence and shortens the life span in senescence-accelerated prone mice. *The Journal of nutritional biochemistry*, 78: 108326. https://doi.org/10.1016/j.jnutbio.2019.108326.

Househam, A. M., Peterson, C. T., Mills, P. J., & Chopra, D. 2017. The Effects of Stress and Meditation on the Immune System, Human Microbiota, and Epigenetics. *Advances in mind-body medicine*, 31(4): 10–25. PMID: 29306937.

Johnsen, P. H., Hilpüsch, F., Cavanagh, J. P., Leikanger, I. S., Kolstad, C., Valle, P. C., & Goll, R. 2018. Faecal microbiota transplantation versus placebo for moderate-to-severe irritable bowel syndrome: a double-blind, randomised, placebo-controlled, parallel-group, single-centre trial. *The lancet. Gastroenterology & hepatology*, 3(1): 17–24. https://doi.org/10.1016/S2468-1253(17)30338-2.

Jung, J. Y., Lee, S. H., Kim, J. M., Park, M. S., Bae, J. W., Hahn, Y., Madsen, E. L., &Jeon, C. O. 2011. Metagenomic analysis of kimchi, a traditional Korean fermented food. *Applied and environmental microbiology*, 77(7): 2264–74. https://doi.org/10.1128/AEM.02157-10.

Kao, D., Roach, B., Silva, M., Beck, P., Rioux, K., Kaplan, G. G., Chang, H. J., Coward, S., Goodman, K. J., Xu, H., Madsen, K., Mason, A., Wong, G. K., Jovel, J., Patterson, J., & Louie, T. 2017. Effect of Oral Capsule- vs Colonoscopy-Delivered Fecal Microbiota Transplantation on Recurrent Clostridium difficile Infection: A Randomized Clinical Trial. *JAMA*, 318(20): 1985–93. https://doi.org/10.1001/jama.2017.17077.

Kim, J. Y., Gum, S. N., Paik, J. K., Lim, H. H., Kim, K. C., Ogasawara, K., Inoue, K., Park, S., Jang, Y., & Lee, J. H. 2008. Effects of nattokinase on blood pressure: a randomized, controlled trial.

Hypertension research: official journal of the Japanese Society of Hypertension, 31(8): 1583–88. https://doi.org/10.1291/hypres.31.1583.

Li, G., Xie, C., Lu, S., Nichols, R. G., Tian, Y., Li, L., Patel, D., Ma, Y., Brocker, C. N., Yan, T., Krausz, K. W., Xiang, R., Gavrilova, O., Patterson, A. D., & Gonzalez, F. J. (2017). Intermittent Fasting Promotes White Adipose Browning and Decreases Obesity by Shaping the Gut Microbiota. *Cell metabolism*, 26(4):672–85.e4. https://doi.org/10.1016/j.cmet.2017.08.019.

Lin, R., Zhang, Y., Chen, L., Qi, Y., He, J., Hu, M., Zhang, Y., Fan, L., Yang, T., Wang, L., Si, M., & Chen, S. 2020. The effects of cigarettes and alcohol on intestinal microbiota in healthy men. *Journal of microbiology (Seoul, Korea)*, 10.1007/s12275-020-0006-7. Advance online publication. https://doi.org/10.1007/s12275-020-0006-7.

Lowe, D. A., Wu, N., Rohdin-Bibby, L., Moore, A. H., Kelly, N., Liu, Y. E., Philip, E., Vittinghoff, E., Heymsfield, S. B., Olgin, J. E., Shepherd, J. A., & Weiss, E. J. 2020. Effects of Time-Restricted Eating on Weight Loss and Other Metabolic Parameters in Women and Men With Overweight and Obesity: The TREAT Randomized Clinical Trial. *JAMA internal medicine*, e204153. Advance online publication. https://doi.org/10.1001/jamainternmed.2020.4153.

Nguyen, N., Zhang, B., Holubar, S. D., Pardi, D. S., & Singh, S. 2019. Treatment and prevention of pouchitis after ileal pouch-anal anastomosis for chronic ulcerative colitis. *The Cochrane database of systematic reviews*, 11(11): CD001176. https://doi.org/10.1002/14651858.CD001176.pub5.

Özkul, C., Yalınay, M., & Karakan, T. 2019. Islamic fasting leads to an increased abundance of Akkermansia muciniphila and Bacteroides fragilis group: A preliminary study on intermittent fasting. *The Turkish journal of gastroenterology: the official journal of Turkish Society of Gastroenterology*, 30(12): 1030–35. https://doi.org/10.5152/tjg.2019.19185.

Paramsothy, S., Kamm, M. A., Kaakoush, N. O., Walsh, A. J., van den Bogaerde, J., Samuel, D., Leong, R., Connor, S., Ng, W., Paramsothy, R., Xuan, W., Lin, E., Mitchell, H. M., & Borody, T. J. 2017.

Multidonor intensive faecal microbiota transplantation for active ulcerative colitis: a randomised placebo-controlled trial. *Lancet (London, England)*, 389(10075): 1218–28. https://doi.org/10.1016/S0140-6736(17)30182-4.

Piercy, K. L., Troiano, R. P., Ballard, R. M., Carlson, S. A., Fulton, J. E., Galuska, D. A., George, S. M., & Olson, R. D. 2018. The Physical Activity Guidelines for Americans. *JAMA*, 320(19): 2020–28. https://doi.org/10.1001/jama.2018.14854.

Sonnenburg, E. D., & Sonnenburg, J. L. 2014. Starving our microbial self: the deleterious consequences of a diet deficient in microbiota-accessible carbohydrates. *Cell metabolism*, 20(5): 779–86. https://doi.org/10.1016/j.cmet.2014.07.003.

Watson, N. F., Badr, M. S., Belenky, G., Bliwise, D. L., Buxton, O. M., Buysse, D., Dinges, D. F., Gangwisch, J., Grandner, M. A., Kushida, C., Malhotra, R. K., Martin, J. L., Patel, S. R., Quan, S. F., & Tasali, E. 2015. Recommended Amount of Sleep for a Healthy Adult: A Joint Consensus Statement of the American Academy of Sleep Medicine and Sleep Research Society. *Sleep*, 38(6): 843–4. https://doi.org/10.5665/sleep.4716.

Yamamoto, S., Sobue, T., Kobayashi, M., Sasaki, S., Tsugane, S., & Japan Public Health Center-Based Prospective Study on Cancer Cardiovascular Diseases Group. 2003. Soy, isoflavones, and breast cancer risk in Japan. *Journal of the National Cancer Institute*, 95(12): 906–13. https://doi.org/10.1093/jnci/95.12.906.

Zhang, C., Zhang, M., Wang, S., Han, R., Cao, Y., Hua, W., Mao, Y., Zhang, X., Pang, X., Wei, C., Zhao, G., Chen, Y., & Zhao, L. 2010. Interactions between gut microbiota, host genetics and diet relevant to development of metabolic syndromes in mice. *The ISME journal*, 4(2): 232–41. https://doi.org/10.1038/ismej.2009.112.

About the Authors

Dr Ujjwal Sonika
Associate Professor
Department of Gastroenterology, GIPMER, Maulana Azad Medical College, University of Delhi

Dr Sonika is a faculty at GB Pant Institute of Postgraduate Medical Education and Research, New Delhi, India. He completed his graduation and post-graduation from the Maulana Azad Medical College, New Delhi and pursued his fellowship in Gastroenterology at the prestigious All India Institute of Medical Sciences, New Delhi. He was the recipient of 'Asia Young Endoscopist Award' fellowship in 2018 at Seoul, South Korea. Dr Sonika has presented his research work at multiple international platforms including APDW. He has published numerous research papers in many National and International journals of repute and has authored various chapters too. He is an expert in the fields of GI endoscopy, acute pancreatitis and inflammatory bowel disease.

Dr Medha Kapoor
Chief Nutrition Consultant, Varsity Skin and Wellness Clinic, New Delhi

Dr. Kapoor is the Co-founder and Chief Nutrition Consultant at Varsity Skin and Wellness Clinic, New Delhi, India. She pursued her

Doctorate in Exercise and Stress Physiology from the Defence Research and Development Organisation, Ministry of Defence, India. She was awarded the 'Young Scientist Award' at the 102nd Indian Science Congress, 2015 by the Government of India. Dr Kapoor has published her research work in many National and International journals of repute and has presented her research work across global platforms including the prestigious International Union of Physiological Sciences (IUPS). She has been associated with the Indian school of Hospitality, Gurgaon, India, as an advisor and adjunct faculty, where she curated and facilitated the semester course curriculum for the subject "Culinary Sociology and Anthropology." Dr Kapoor specializes in the management and prevention of lifestyle diseases such as Obesity, Diabetes, Hypertension, Dyslipidemia, Thyroid disorders, Polycystic Ovarian Disease, Heart Diseases and Certain types of Cancers. She has helped thousands of people with the aforementioned disorders improve their quality of life by incorporating tailor-made dietary and exercise based recommendations. She conducts wellness sessions for Corporates and Educational Institutions and runs lifestyle modification programs for individuals.

INDEX

A

B

E

F

G

H

I

J

K

L

M

N

O

P

R

S